GREENHOUSE GAS EMISSIONS FROM ECOTECHNOLOGIES FOR WASTEWATER TREATMENT

Juan Pablo Silva Vinasco

GREENHOUSE GAS EMISSIONS FROM ECOTECHNOLOGIES FOR WASTEWATER TREATMENT

Thesis
submitted in fulfilment of the requirements of
the Academic Board of Wageningen University and
the Academic Board of the IHE Delft Institute for Water Education
for the degree of doctor
to be defended in public
on Monday, 14 September 2020 at 3:30 p.m.
in Delft, the Netherlands

by Juan Pablo Silva Vinasco
Born in Cali, Colombia

CRC Press/Balkema is an imprint of the Taylor & Francis Group, an informa business

Published by:
CRC Press/Balkema
Schipholweg 107C, 2316 XC, Leiden, the Netherlands
Pub.NL@taylorandfrancis.com
www.crcpress.com – www.taylorandfrancis.com

ISBN: 978-0-367-67382-6 (Taylor & Francis Group)
ISBN: 978-94-6395-449-5 (Wageningen University)
DOI: https://doi.org/10.18174/525933

DEDICATION

To the memory of Angel and Eufemia, my beloved father and mother, who died without seeing this great achievement.

To Elida and Diana, who have given me their support and love

ACKNOWLEDGEMENTS

The author is grateful to the European research project SWITCH (Sustainable Urban Water Management Improves Tomorrow's City's Health) for financially supporting this research under the 6th Framework Programme, contributing to the thematic priority area 'Global Change and Ecosystems' [1.1.6.3] Contract Nr. 018530-2.

I am specially indebted to my promotor, Professor Dr. Huub Gijzen for the valuable academic advice and support that he gave me in the most difficult moments of this PhD study. His dedicated attention to the quality of my work and continuous encouragement teach me the necessity of perseverance.

To my co-promotor Dr. Henk Lubberding, who spent many hours discussing and giving suggestions along these years. I want to acknowledge my colleague Dr. Miguel Peña from Universidad del Valle for his support not only as co-supervisor but also as a friend.

I gratefully acknowledge the Universidad del Valle for providing all possible support to allow me to complete this work. Many thanks to colleagues and friends of the School of Natural Resources and Environmental Engineering for their support.

To my friends Carlos Madera, Johnny Rojas, Miguel Peña, Arlex Sanchez, Carlos Vélez and Mario Pérez for all the special moments we have shared and for teaching me that life goes beyond a PhD.

Thanks go also to Acuavalle, and specially to the laboratory staff Yimmer Vélez and Haidér Pérez for their valuable support to this investigation.

To my family, this challenge belongs to them.

To all those who asked me daily when this PhD would end.

Table of contents

List of Figures

List of Tables

List of abbreviations and acronyms

AFP	algae facultative ponds
AOB	ammonia-oxidizing bacteria
AS	activated sludge
AP	anaerobic pond
b.p.	barometric pressure
BNR	biological nutrient removal
BOD	biological oxygen demand
C/N	carbon/nitrogen ratio
CFC's	chlorofluorocarbons
CO	carbon monoxide
COD	chemical oxygen demand
$kgCOD_{rem}$	kilograms chemical oxygen demand removed
COD/N	chemical oxygen demand/nitrogen ratio
CO_2	carbon dioxide
CH_4	methane
CW	constructed wetland
DBP	duckweed-based ponds
DO	dissolved oxygen
EFs	Emission factors to calculate emissions
EWWT	Ecotechnologies for wastewater treatment
FWS	free-water surface wetland
GHG	greenhouse gas
$GtCO_2$	gigatonnes carbon dioxide
GWP	global warming potential
H_2	hydrogen
HRT	hydraulic retention time
HP	hyacinth ponds
HRalP	high-rate algal pond
HRAP	high rate anaerobic ponds
HSSF	horizontal subsurface wetland
IPCC	intergovernmental panel climate change
km^2	square kilometre
MASL	meters above sea level
N_2	nitrogen
NH_4^+	ammonium
NO_3^-	nitrate
NO_2^-	nitrite
NO_x	nitrogen oxides
N_2O	nitrous oxide

N_2O-N	nitrous oxide emitted relative to the nitrogen load
NO	nitric oxide
TN	Total nitrogen
OM	organic matter
O_3	ozone
OH	hydroxyl radical
ORP	oxidation-reduction potential
Pg	petagrams or 10^{15} grams
ppmv	parts per million in volume
ppbv	parts per billion in volume
R^2	regression coefficient
R^2_{adj}	regression coefficient adjusted
$R_{Pearson}$	Pearson regression coefficient
SFPs	secondary facultative ponds
SWITCH	Sustainable Urban Water Management Improves Tomorrow's City's Health
TN	total nitrogen
TKN	total Kjeldahl nitrogen
TSI	Carlson Trophic State Index
TOC	total organic carbon
TP	total phosphorus
TSS	total suspended solids
UASB	up-flow anaerobic sludge blanket
VSS	volatile suspended solids
VSSW	vertical subsurface wetland
$W.m^{-2}$	watts per square meter
WSP	wastewater stabilization ponds
WWT	wastewater treatment
yr	year

Chapter 1

General Introduction

1.1 THE GREENHOUSE EFFECT, GLOBAL WARMING AND CLIMATE CHANGE

The atmosphere is layer of gases surrounding the planet Earth. It contains three primary gases, nitrogen (78.09%), oxygen (20.95%), and argon (0.93%). Furthermore, the atmosphere contains trace gases like carbon dioxide (CO_2), methane (CH_4), carbon monoxide (CO), nitrous oxide (N_2O), nitric oxide (NO), chlorofluorocarbons (CFCs), water vapour (H_2O) and ozone (O_3).

These trace gases are known as greenhouse gases (GHGs) because they contribute to the greenhouse effect. When radiant energy from the sun come to the Earth's atmosphere, most of energy is reflected back to space. While clouds and aerosols absorb some portion, a smaller portion is also absorbed by greenhouse gases. GHGs absorb and reradiate downward a large fraction of infrared wave lengths (i.e., 8 to 12 µm), which leads to warming of the Earth's surface. Without this heat trapping by the GHGs in the atmosphere, the surface of the Earth would be about 20°C colder than it actually is (Forster et al., 2007). However, increasing GHGs concentrations on the atmosphere may to lead to global warming and climate change.

Growing atmospheric concentrations of GHGs have been caused by fossil fuel combustion and other human activities such as farming, manufacturing, waste disposal and deforestation **(Figure 1.1)**. Of the cumulative human CO_2 atmospheric emissions between 1750 and 2011 (2040±310 $GtCO_2$), half has taken place since 1974 (IPCC, 2014). The pre-industrial levels (prior to 1750) of CO_2, CH_4, and N_2O had risen by 2011 to 397.7 ppmv, 1833 ppbv, and 327 ppbv, respectively, an increase of approximately 43%, 154%, and 21% respectively (Tarasova et al., (2016).

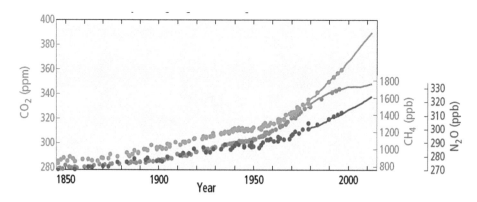

Figure 1.1 Trends of greenhouse gas concentrations since the preindustrial period.
Source: (IPCC, 2014)

1.2 OVERVIEW OF MAJOR GREENHOUSE GASES

1.2.1 CARBON DIOXIDE

The IPCC (2013) states that the chief causes for the worldwide greenhouse effect are CO_2 and water vapour. When CO_2 has been released into the atmosphere, it cannot be eliminated by chemical breakdown. Instead, it is redistributed to other carbon reservoirs such as oceans and freshwater systems. Ravindranath and Sathaye (2002) state that although the majority of the CO_2 emissions are abstracted within approximately 100 years, a portion is more or less permanent and stays in the atmosphere for millenniums.

Due to human activities the atmospheric CO_2 concentration has increased by 40% from about 280 ppmv in the pre-industrial era (before 1750) to 398 ppmv in 2011 (Tarasova *et al.*, 2016). These changes were even more noteworthy in the past decade (2000-2010) where CO_2 concentrations increased on average by about 2.1 ppmv. yr^{-1} (Prather *et al.*, 2012). Thus, IPCC (2013) mentioned that the CO_2 has had an average growth rate in terms of radiative force of 0.27 (0.16 to 0.30) $W.m^{-2}$ per decade.

Anthropogenic CO_2 emissions into the atmosphere were 555±85 PgC between 1750-2011 (**Table 1.1**); of this amount fossil fuel combustion and cement production contributed 375±30 PgC and land use (including deforestation, afforestation, and reforestation) contributed 180±80 PgC (IPCC, 2014). In addition, during this time-period the oceans sequestered 155±30 PgC, whereas vegetation biomass and soils not affected by land use change sequestered 160±90 PgC from the atmosphere (Ciais *et al.*, 2014). Thus, about half of the emissions since 1750 (240±10 PgC) have remained in the atmosphere (Ciais *et al.*, 2014).

Table 1.1 CO_2 accumulation in the atmosphere from 1750 to 2011.

	1750-2011
CO_2 source/sink	cumulative PgC
Atmospheric increase	240 ± 10
Fossil fuel combustion and cement production	375 ± 30
Ocean-to-atmosphere flux	-155 ± 30
Land-to-atmosphere flux	30 ±45
Partitioned as follows	
Net land use change	180 ± 80
Residual land sink	-160 ±90

Source: (Ciais *et al.*, 2014).

1.2.2 Methane

Methane is currently the most abundant non-CO_2 greenhouse in the atmosphere. Methane concentration in the atmosphere was 722±25 ppbv in 1750, increasing to 1,803±2 ppbv in 2011 (IPCC, 2013). From the 1980s until about 1992, atmospheric methane concentration rose sharply by about 12 ppbv.yr^{-1}. After that concentrations decreased by about 3ppbv.yr^{-1} for about a decade until stabilization was reached in 1999.

Because the emissions reported for the 1990s were highly variable due to uncertainty in estimates of anthropogenic sources of methane, the explanation to the steady behavior of the concentrations is not well known. However, from 2007 to 2011 CH_4 were rising again by about 6 ppbv.yr^{-1} (Dlugokencky et al., 2009; Nisbet et al., 2014). According to EPA (2012), the recovery and use of methane from wastes, and practices more efficient in farming (ruminants) could have contributed to the stabilization of the concentrations of this gas in the atmosphere

Although global CH_4 emissions are only 4% of the global CO_2, atmospheric CH_4 has contributed about 20% (~ 0.49 W.m^{-2}) of the additional radiative force accumulated in the lower atmosphere since 1750 (Ciais et al., 2014). This is because the global warming potential (GWP) of methane compared to CO_2 is 28:1 on a 100-year horizon as used by the International Panel on Climate Change (Myhre et al., 2013).

The global methane emission (bottom-up) budget for the decade 2000-2009 is 678 TgCH$_4$ yr^{-1} (Hartmann et al., 2013;Ciais et al., 2014). The main natural sources of CH_4 (**Figure 1.2**) are wetlands (177 to 284 TgCH$_4$.yr^{-1}). Most wetland emissions (70%) come from the tropics and are enhanced during warm, wet periods, and high water tables. Smaller amounts of CH_4 are emitted from oceans, and by termites (combined contribution of 65 TgCH4 yr^{-1}) (EPA, 2010). Anthropogenic emissions are about 60% of total emissions (Saunois et al., 2016). Agriculture (rice fields, ruminants, and biomass burning) and fossil fuel exploitation together account for about 256 TgCH$_4$.yr^{-1}, while smaller emissions come from waste treatment (land fill, manure and sewage) at 75 TgCH4 yr^{-1} (Ciais et al., 2014). Methane production from ruminant livestock is estimated at between 87 and 94 TgCH$_4$.yr^{-1} (EDGAR database, 2009).

The methane sinks contribute to regulating the concentrations of this gas in the atmosphere. About 90% of the CH_4 emitted (509 to 764 TgCH$_4$.yr^{-1}) into the atmosphere is destroyed by photochemical oxidation with OH$^.$ radicals (Young et al., 2013), occurring mostly in the troposphere (Ghosh et al., 2015). Another minor sink process for methane is the stratospheric loss by reaction with radical OH$^.$, Cl$^.$ and O (^{1}D) (excited oxygen atoms), resulting in a combined loss rate of 76 TgCH$_4$.yr^{-1}. Finally, the process related to methane oxidation by methanotrophic bacteria in upland soils, natural wetlands, and rice paddies, removes about 9 to 47 TgCH$_4$.yr^{-1} (Spahni et al., 2011).

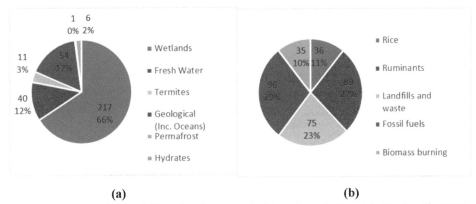

Figure 1.2 Sources of natural (a) and anthropogenic (b) methane in the global budget for 2000-2009. Source: (Ciais *et al.*, 2014). The data are reported in TgCH$_4$.yr^{-1} and the percentage represents each source contribution.

1.2.3 Nitrous oxide

Nitrous oxide is an important greenhouse gas contributing to global warming and to the depletion of stratospheric ozone (Tallec *et al.*, 2008; Ravishankara *et al.*, 2009). Worldwide the average atmospheric concentrations of N$_2$O increased by 20% from 270 ppbv in 1750 to 324.2 ppbv in 2011 (Prather *et al.*, 2012). The problem that arises with this increase in atmospheric N$_2$O concentrations is that the atmospheric lifetime for nitrous oxide is about 120 years and its global warming potential is 296 relatives to CO$_2$ over a 100-year time horizon (Sovik and Klove, 2007; Mander *et al.*, 2014). Since 2011, N$_2$O has become the third largest contributor to the radiative force (0.14 to 0.20 W.m^{-2}) (Myhre *et al.*, 2013).

Between 2006 and 2011 the global average N$_2$O emissions amounted to 17.9 (ranging from 8.1 to 30.7) TgN$_2$O-N.yr^{-1} (Prather *et al.*, 2012; Ciais *et al.*, 2014). Natural sources such as oceans, soil processes, forest and bush fires emitted about 60% of total N$_2$O (11 TgN$_2$O-N.yr^{-1}) (Mosier, 1998; Kroeze *et al.*, 1999; Barton and Atwater, 2002; Crutzen *et al.*, 2007). Anthropogenic average N$_2$O emissions account for 6.9 (ranging from 2.7 to 11.1) Tg N$_2$O.yr^{-1} and arise primarily from agricultural activities (4.1 TgN$_2$O-N.yr^{-1}) (**Figure 1.3**) and industrial processes including fossil fuel combustion (0.7 TgN$_2$O.yr^{-1}) (Montzka *et al.*, 2011). There are a variety of sources of N$_2$O in agricultural systems, which include synthetic fertilizers, animal manure (urine and faeces), crop residues returned to the field after harvest and human sewage sludge application. Of concern in recent years is increased fertilizer use, causing over-fertilization which stimulates the potential increase in N$_2$O emissions (Paustian *et al.*, 2016).

The main sink of N$_2$O is its photo-dissociation into N$_2$ and O$_2$ which occurs at an altitude above 30 km in the stratosphere (UNEP, 2013). According to Montzka *et al.* (2011), less than 1% of atmospheric N$_2$O is removed annually from the atmosphere, primarily by photolysis and oxidative reactions in the stratosphere. Nitrous oxide is also converted (and produced) by denitrification in nearly anoxic environments and, possibly destroyed in forest soils under drought (Goldberg and Gebauer, 2009; UNEP, 2013).

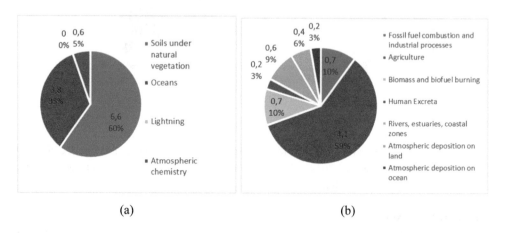

(a)   (b)

Figure 1.3 Sources of natural (a) and anthropogenic (b), nitrous oxide for 2006-2011. (Prather *et al.*, 2012; Ciais *et al.*, 2014). The data are reported in TgCH$_4$ year^{-1} and the percentage represents each source contribution.

1.3 GHG FROM THE WASTE MANAGEMENT SECTOR

Waste management and treatment activities are sources of GHG emissions (EPA, 2016). The waste sector is the third largest contributor to global emissions of non-CO$_2$ GHG with the largest GHG source being landfill methane, followed by wastewater CH$_4$ and N$_2$O (Yusuf *et al.*, 2012). On the other hand, aerobic composting and incineration of waste containing fossil carbon such as plastics or synthetic textiles prevent methane production and its subsequent release into the atmosphere (Bogner *et al.*, 2008).

The quantity of methane from waste is projected to be 1,276.3 MtCO$_{2eq}$ annually (EPA, 2006). This emission accounts for 20.6% of the anthropogenic methane emissions. The emissions from landfilling of solid waste (59.1%) and wastewater (40.8%) are the two largest sources of emissions in this sector (Karakurt *et al.*, 2012). Methane emissions from landfilling of solid waste dropped from 761 MtCO$_{2eq}$ in 1990 to 730 MtCO$_{2eq}$ in 2000 and rose to 761 MtCO$_{2eq}$ in 2010 (**Figure 1.4**). It is projected to reach 817 MtCO$_{2eq}$ by 2020 (Bogner *et al.*, 2007).

Abatement options to prevent GHG resulting from landfill involve several issues. An example is the design of engineered landfills to collect and to recover the biogas to generate directly electricity, direct gas use (injection into natural gas pipelines), powering fuel cells, or compression into liquid fuel (Bogner *et al.*, 2007; Yusuf *et al.*, 2012). Further, waste management practices can be changed to reduce waste disposal (waste minimization) at landfills by adding composting and programs for recycling and reuse (waste diversion). Incineration is another possible consideration (Yusuf *et al.*, 2012).

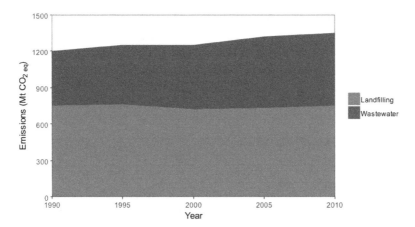

Figure 1.4 Emission trends in waste management sector. Source: (Karakurt *et al.*, 2012)

GHG emissions from the wastewater treatment sector represents 3-4% of total GHG emissions. However, the contribution of the wastewater sector to GHG emissions may be underestimated. This is due to large uncertainties with respect to direct emissions, indirect emissions and the availability and quality of annual data for the wastewater sector (Bogner *et al.*, 2007). Methane and nitrous oxide emissions from conventional wastewater treatment operating in most developed countries are low and circumstantial (EPA, 2012). By contrast, in developing countries, in areas where there are no sewer systems, wastewater treatment is insufficient or based on anaerobic systems such as latrines, open sewers, or stabilization ponds, methane and nitrous oxide emissions are higher and more uncontrolled than conventional systems (EPA, 2006; Bogner *et al.*, 2008; Karakurt *et al.*, 2012). Further, in most developing countries large volumes of wastewaters will not receive any treatment, leading to oxygen depletion, and corresponding CH_4 and NO_x emissions from receiving water bodies.

Between 1990 and 2025, global CH_4 and N_2O emissions from wastewater are estimated to increase from 352 to 477 $MtCO_{2eq}$ and from 82 to 100 $MtCO_{2eq}$, respectively (EPA, 2012). This growth in GHG emissions will come from developing countries of East and South Asia, the Middle East, the Caribbean, and Central and South America, mainly due to population increase (Bogner *et al.*, 2007). As long as populations continue to grow significantly without large-scale advances in wastewater treatment, these areas will continue to have a major influence on the upward trend in GHG emissions.

The reduction in GHG emissions from wastewater can be achieved through improved wastewater treatment practices. In systems under anaerobic conditions the generated CH_4 can be captured and used as an energy source either on-site in the wastewater facility, or off-site (Bogner *et al.*, 2008). This energy recovery is more viable when applying high-rate anaerobic processes for the treatment of liquid effluents (concentrated sewage) (Gijzen, 2002). The commercial exploitation of the generated biogas is important, because it positively affects the overall energy balance of the process and replaces an equivalent amount of non-renewable energy and greenhouse gas emissions (Gijzen, 2001). Alternatively, when the biogas yield is insufficient to provide energy recovery, it can be flared, which converts CH_4 to CO_2, with a much lower global warming potential.

1.4 GHG PRODUCTION IN WWT

Wastewater treatment generates greenhouse gases, particularly CO_2, CH_4, and N_2O (Bogner *et al.*, 2007; Foley *et al.*, 2011; Daelman *et al.*, 2013; EPA, 2016). These gases are produced by biochemical transformations during the sewage collection and wastewater treatment.

CO_2 is produced in both aerobic and anaerobic biological wastewater treatment. During aerobic treatment, organic matter is oxidized into CO_2 and other metabolites by heterotrophic bacteria while in anaerobic treatment, the organic compounds are transformed into biogas, a gas mixture of CO_2 and CH_4 (30-40% and 60-70% v/v, respectively). An additional source of CO_2 emissions in WWT is related to the alkalinity depletion in the bicarbonate form (HCO_3^-) at near-neutral pH (Das, 2011) . In addition, CO_2 is also emitted during the production of energy required for the plant operation (Campos *et al.*, 2016).

Methane production is attributed to the anaerobic transformation of complex macromolecules present in sewage (**Figure 1.5**). According to Gujer and Zehnder (1983), the anaerobic digestion involves four big stages : *hydrolysis, acidogenesis, acetogenesis* and *methanogenesis*. During *hydrolysis*, complex organic compounds such as carbohydrates, proteins and lipids are broken down into simple sugars, amino acids, and fatty acids by fermentative bacteria. During *acidogenesis*, fermentative bacteria convert the products from the previous stage into short-chain acids such as acetic, propionic, formic, lactic, and butyric acids; in addition, alcohols and ketones, CO_2 and H_2 are produced. *Acetogenesis* involves the conversion of simple molecules from acidogenesis into acetic acid and acetate, which are key substrates for methanogens in the final stage of anaerobic digestion. Finally, in the methanogenesis, methane is produced from acetate or from the reduction of carbon dioxide by acetotrophic and hydrogenotrophic bacteria, respectively.

The N_2O emitted directly from wastewater treatment can be an intermediate product of both nitrification and denitrification process but has typically been associated with denitrification (**Figure 1.6**) (Law *et al.*, 2012; EPA, 2016). During the nitrification, the absence or a low concentration of oxygen (<0.5 mg.L^{-1}) limits the total oxydation of ammonium (NH$_4^+$) to nitrate (NO$_3^-$) nitrate and thus N$_2$O is produced as an intermediary because incomplete oxidation of NH$_2$OH (Guo *et al.*, 2018). Denitrification is the stepwise reduction of NO$_3^-$ to N$_2$ by specialized bacteria (denitrifyers), thereby generating intermediary products such as NO$_2^-$, NO and N$_2$O (Kampschreur *et al.*, 2009; Law *et al.*, 2012; Daelman *et al.*, 2013). N$_2$O formation during denitrification occurs when the WWT is operated at low pH values, limited dissolved oxygen and low C/N (Thomson *et al.*, 2012).

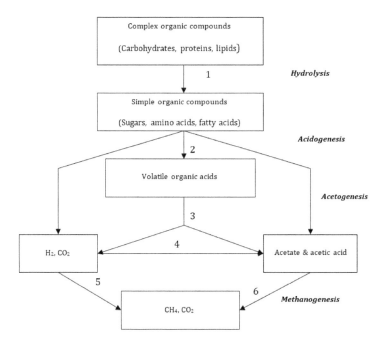

Figure 1.5 Reactive scheme for the anaerobic digestion of polymeric materials. Adapted from (Gujer and Zehnder, 1983). Numbers indicate the bacterial groups involved: 1. Hydrolytic and fermentative bacteria, 2. Acidogenic bacteria, 3. Acetogenic bacteria, 4. Homo-acetogenic bacteria, 5. Acetoclastic methanogens, 6. Hydrogenotrophic methanogens.

Indirect emissions of N$_2$O from wastewater treatment can be caused by ammonia volatilization during the treatment. Indirect emissions of N$_2$O from wastewater treatment can be caused by ammonia volatilization during the treatment. However, according to Barton and Atwater (2002) ammonia from WWTS is ignored as N$_2$O precursor in other processes. For instance, when ammonia is released into the atmosphere a fraction of this is oxidized to NO$_x$ which react to produce ammonium-containing aerosols, such as (NH$_4$)$_2$SO$_4$ and NH$_4$NO$_3$. Once formed these aerosols, are transported to both land surface and water-bodies via dry or wet deposition. Thus, this reactive nitrogen will be available again to stimulate nitrification/denitrification process which could lead to subsequent production and emission of N_2O (Barton and Atwater, 2002).

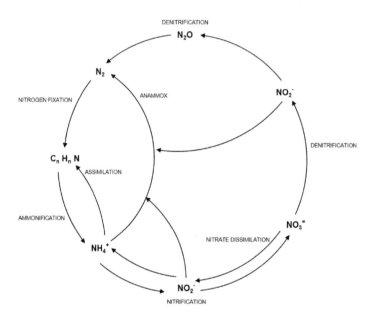

Figure 1.6 Nitrogen transformation in wastewater treatment. Adapted from (Thomson *et al.* 2012)

1.5 ECOTECHNOLOGIES FOR WASTEWATER TREATMENT

Ecotechnologies for wastewater treatment (EWWT) combine ecological principles of natural systems with engineering principles to improve removal of organic carbon, nutrients and pathogenic microorganisms from wastewater. EWWT are mainly solar-based systems, which minimise the dependence on external energy. Similar to the situation in nature, in EWWT a diversity of living organisms (biotic) interact with the non-living components (abiotic) to boost biochemical transformations of organic matter and nutrients into more stable compounds. The main principles of engineering applied to EWWT are those related to improve hydraulic, kinetic, and mass transfer. This may be carried out designing adequate entrance and exit structures, selecting the plant species in EWWT, recycling to increase biomass retention, adding heat, chemicals, and air among others.

Examples of EWWT include algae facultative ponds (AFPs), anaerobic ponds (APs), duckweed-based ponds (DBPs), and constructed wetlands (CWs). These ecotechnologies

(**Table 1.2**) are appropriate and economically feasible for many developing countries, due to the following aspects: sufficient land availability, favourable climate, simple operation, little or no equipment required (Von Sperling and Chernicharo, 2005; Ho *et al.*, 2017), effluent reuse and recycling, nutrient recovery, biomass production for energy, and possible combination or linkage to other productive applications such as agriculture or aquaculture.

Considerable research all over the world has demonstrated the advantages of EWWT compared to conventional wastewater treatment systems. This includes research on waste stabilization ponds (Arthur, 1983; Peña *et al.*, 2002; Mara, 2005; Verbyla *et al.*, 2016), constructed wetlands (Rousseau *et al.*, 2004; Kadlec *et al.*, 2005; Vymazal, 2007; Vymazal and Březinová, 2016), and duckweed ponds (Alaerts *et al.*, 1996; Nhapi *et al.*, 2003; Zimmo *et al.*, 2003; Caicedo, 2005; Nhapi and Gijzen, 2005; El-Shafai *et al.*, 2007; Sekomo *et al.*, 2012; Sims *et al.*, 2013; Verma and Suthar, 2016).

However, EWWT operation may lead to greenhouse gas emissions and odour generation (Crites *et al.*, 1995; Van der Steen *et al.*, 2003; Shilton and Walmsey, 2005; Hernandez-Paniagua *et al.*, 2014; Mander *et al.*, 2014; Paredes *et al.*, 2015; Glaz *et al.*, 2016). CO_2 and CH_4 emissions have been measured in anaerobic ponds (Toprak, 1995; Picot *et al.*, 2003; Wang *et al.*, 2011; Konaté *et al.*, 2013; Paredes *et al.*, 2015), facultative ponds (Stadmark and Leonardson, 2005; Hernandez-Paniagua *et al.*, 2014; Detweiler *et al.*, 2014; Glaz *et al.*, 2016), and constructed wetlands (Tanner *et al.*, 1997; Fey *et al.*, 1999; Johansson *et al.*, 2003; Johansson *et al.*, 2004; Mander *et al.*, 2005; Teiter and Mander, 2005; Liikanen *et al.*, 2006; Sovik *et al.*, 2006; Gui *et al.*, 2007; Sovik and Klove, 2007; De Klein and Van der Werf, 2014; Wu *et al.*, 2016). Therefore, there is a risk that water pollution control by EWWT can turn into an atmosphere pollution problem.

So far, limited information is available on the fate of GHG production or consumption in EWWT operated under tropical conditions. In addition, there is lack of understanding of CO_2, CH_4, and N_2O production in function of environmental parameters and wastewater characteristics. The tropical conditions are characterized by high temperatures, long and stable photoperiods, photosynthetic activity, and dynamics in DO and pH patterns which may all influence GHG dynamics differently compared to temperate conditions. Because of the global concern about the possible effects of human activities on global warming, further studies of GHGs from EWWT under different environmental and operational conditions need to be developed.

1.6 GHG EMISSIONS FROM WWT

1.6.1 GHG emissions from conventional wastewater treatment

In developed countries, the most common methods of municipal wastewater treatment processes are centralized aerobic treatment processes (Das, 2011), i.e. activated sludge (AS).

Table 1.2 Characteristics of ecotechnologies for wastewater treatment

Type	Characteristic	Design depth (m)	HRT (days)	Organic loading	Removal (%) BOD	Removal (%) TN	Removal (%) TP	Advantages	Disadvantages
	Anaerobic	2-5	1-2	>100 [a]	> 60	----	----	• Simplicity • Low cost • High efficiency • Robustness	• Odour release • Large land area • GHG i.e. CH_4 and CO_2
WSPs	Facultative	1-1.5	5	100-400 [b]	80-95[c]	<70[c]	40-50[c]	• Simplicity • Low cost • High efficiency • Robustness • Effective reduction of coliforms	• Effluent quality (algae presence) • Large land area • Mosquito breeding
	Duckweed	0.4-1.5[d,e]	7-15	100-184[b,c]	60-90	54[d]	61-74[d]	• Low energy • Duckweed controls growth algae and odour. • Resource recovery by harvesting and utilizing the biomass.	• Large land area • Growth reductions at temperatures below 15 ° C. • Less efficient for pathogen removal • Large land area

a) Volumetric loading in $gBOD.m^{-3}.d^{-1}$;   b) Surface BOD loading $kg.ha^{-1}.d^{-1}$ c) Mara (2005) and Pearson et al. (1996)   d) Zimmo et al. (2003)
e) Caicedo (2005)

Table 1.2 Characteristics of ecotechnologies for wastewater treatment (cont.)

Type	Characteristic	Design depth (m)	HRT (days)	Organic Loading	Removal (%) BOD	TN	TP	Advantages	Disadvantages
	FWS	0.3-0.4[a]	2-5 BOD[a] 7-14 for N[a]	<70	74-96	45-58	34-50	• Effective in removal of organics through microbial degradation and removal of suspended solids through filtration and sedimentation. • Stormwater runoff and mine drainage waters • Tertiary treatment.	• Odour, Mosquitos • Large land area. • FWS CWs provide limited removal of phosphorus
CWs	HSSF	0.3-0.8[a]	3-4 for BOD[a] 6-10 for N[a]	<70	75	30-45[b]	50[b]	• Prevents odour • Effective in removal of organics, suspended solids, microbial pollution and heavy metals. • Stormwater runoff and mine drainage waters. • Low operation and maintenance costs.	• Not appropriate systems for ammonia or phosphorus removal. • Clogging • Large land area. • Limits nitrification • High cost of bed (gravel)
	VSSW	1.4[b]	4-15 for N and P[b]	[b]Area: 1-3 m^2 PE^{-1}	90	43[b]	56[b]	• Oxygen transfer • Prevents odour • Nitrogen removal	• Clogging • High costs of sand or gravel • Large land area

a)Lens *et al.* (2001); b) Average daily load per PE (g/d) 60;

Activated sludge is a conventional biological process, capable of removing organic matter and nutrients. This process uses the microbial community suspended in the wastewater to metabolize the biodegradable organic and inorganic components (Law et al., 2012). However, despite its extensive use, the AS system is not the most sustainable method due to its high energy consumption and low energy recovery, resulting in high costs and an environmental footprint i.e. GHG emissions (Gijzen, 2002; Verstraete and Vlaeminck, 2011). In addition, the operation of AS produces direct greenhouse gas emissions i.e. CO_2, CH_4 and N_2O from the biological processes.

Two processes generate the N_2O emitted from conventional wastewater treatment: nitrification and denitrification. These processes are used to remove nitrogen compounds from wastewater (**Figure 1.6**). Nitrous oxide is emitted in the centralized aerobic treatment predominantly in the aerobic tank (Czepiel et al., 1995; Ahn et al., 2010; Foley et al., 2015). This could be explained because nitrite accumulation leads to the formation of N_2O in aerobic zones as a result of low oxygen levels, sudden changes in ammonium load, and higher temperatures (Wunderlin et al., 2012; Foley et al., 2015).

In a study in different countries has been reported that N_2O emissions from AS systems show a large variation (**Table 1.3**), ranging from 0 to 25% of the nitrogen load entering into the influent (Ahn et al., 2010; Foley et al., 2015). This suggests that N_2O emissions are highly variable among different plants, and even with the same plant during different seasons of the year or through the day depending of its characteristics or operation (Foley et al., 2015). Furthermore, the different measurement protocols of N_2O emissions could be a contributing factor to the variability in emissions reported (Law et al., 2012).

The main sources of methane from AS wastewater treatment are the primary sludge thickener, the exhaust gas of the cogeneration plant, the buffer tank for the digested sludge, and the storage tank for the dewatered sludge (Daelman et al., 2012). The latter two contribute substantially to methane emission (72%) while the remaining emissions (28%) come from the biological reactor and can be mainly attributed to the CH_4 dissolved in the wastewater (Daelman et al., 2012).

Reports on CH_4 emissions from conventional wastewater treatment are scarce (**Table 1.3**). Two studies reported that conventional wastewater treatment emitted 1.6 and 0.8 g CH_4. $(kgCOD_{influent})^{-1}$ (Czepiel et al., 1993; Wang et al., 2011). Daelman et al. (2012) reported that about 1% of the incoming chemical oxygen demand (COD) to a wastewater treatment plant was emitted as methane.

Research work by Hwang (2016) also showed that most of the CH_4 emissions are closely related to processes occurring primarily from the inlet works and anaerobic conditions during wastewater transport. In a study by Foley et al. (2015), the highest methane emissions were observed from the aeration tank, which probably were due to the presence of micro-bubbles of methane suspended in the water (about 25%), which probably was generated in the sewer system. However, the major emissions of CH_4 measured in conventional wastewater treatment were from the sludge handling and storage sites such as the anaerobic digester (in total ~50%) (Foley *et al.*, 2015).

Table 1.3 Methane and nitrous oxide emission factors reported for several centralized aerobic treatment processes.

Type of plant	N_2O emission (% of N-influent)	CH_4 emission (kg CH_4/kgCOD$_{influent}$)	References
Activated sludge	0.035-0.05	1.6	Czepiel *et al.* (1993) and Czepiel *et al.* (1995)
Partial nitritation-anammox sequencing batch reactor	0.4-0.6	n.d	Joss *et al.* (2009)
Nitritation anammox	2.3	n.d.	Kampschreur *et al.* (2009)
12 BNR plants	0.03-2.59	n.d.	Ahn *et al.* (2010)
Activated sludge, plug flow in France, Netherlands, USA and Australia	0.01-11.2	<0.0004-0.048	Foley *et al.* (2015)

n.d. = no data

1.6.2 GHG emissions from ecotechnologies for wastewater treatment
1.6.2.1 Greenhouse gas emissions from anaerobic ponds

Greenhouse gas emissions from anaerobic ponds treating wastewater are reported in **Table 1.4**. All 9 studies of APs reported CH_4 emission data, while only 2 provided data on CO_2 (Toprak, 1995; Hernandez-Paniagua *et al.*, 2014). None of the studies provided N_2O data from APs.

The references shown in the table present information on APs treating different wastewater types including municipal (Toprak, 1995; Picot *et al.*, 2003; Wang *et al.*, 2011; Konaté *et al.*, 2013; Paredes *et al.*, 2015) and livestock, palm oil and tapioca industries (De Sutter and Ham, 2005; Yacob *et al.*, 2005; Hasanudin *et al.*, 2006). Regarding climate conditions, there have been studies on CH_4 and CO_2 release under tropical (Yacob *et al.*, 2005), subtropical (Toprak, 1995 ; Picot *et al.*, 2003; Hernandez-Paniagua *et al.*, 2014; Paredes *et al.*, 2015), and temperate conditions (De Sutter and Ham, 2005; Hasanudin *et al.*, 2006). In all studies, there were large variations on CH_4 and CO_2 measured in APs. The CH_4 and CO_2 emissions inAPs varied from

5,600 to 1,200,000 and 200 to 370,000 mg. m^{-2} d^{-1}, respectively. The main reasons for these large variances relate to differences in local climate conditions, wastewater type and COD loading rates (Toprak, 1995; Picot et al., 2003; Yacob et al., 2005; Konaté et al., 2013).

Table 1.4 Emissions of greenhouse gases from WSPs as reported in literature

Location	Wastewater source	Measurement condition	CO_2[a] $g.m^{-2}.d^{-1}$	CH_4[a] $g.m^{-2}.d^{-1}$	N_2O[a] $g.m^{-2}.d^{-1}$	References
Anaerobic ponds						
Portugal	Municipal	Subtropical	5.5-17.6	16-32	n.d.	Toprak (1995)
France	Municipal	Mediterranean	n.d.	9-77	n.d.	Picot et al. (2003)
USA	Livestock facilities	Temperate	n.d.	60-210	n.d.	De Sutter and Ham (2005)
	Palm oil effluent	Tropical	n.d	257-1,233	n.d.	Yacob et al. (2005)
	Tapioca WW	Temperate	n.d	396-761	n.d.	Hasanudin et al. (2006)
	Pig farming	Temperate	5.6-366	18-442	n.d.	Craggs et al. (2008)
	Municipal	Sahelian	n.d.	20-135	n.d.	Konaté et al. (2013)
	Agricultural WW	Subtropical	0.2	7±1	n.d.	Hernandez-Paniagua et al. (2014)
Mexico	Municipal	Subtropical	n.d.	5.7-59	n.d.	Paredes et al. (2015)
Algal facultative ponds						
Sweden	Municipal	Spring and summer	n.d.	-0.4-1.7	-0.009-0.04	Johansson et al. (2003);Johansson et al. (2004)
India	Municipal	Subtropical	n.d.	n.d..	0.0-0.0005	Singh et al. (2005)
Sweden	Municipal	Temperate	n.d.	0.011-0.97	n.d..	Stadmark and Leonardson (2005)
Mexico	Agricultural WW	Subtropical	0.2-1.0	0.6±0.40	$0.12x10^{-4}$-$0.95x10^{-4}$	Hernandez-Paniagua et al. (2014)
USA	Municipal	Temperate	n..d.	3.-7.4	n.d.	Detweiler et al. (2014)
Canada-Australia	Municipal	Temperate	-0.8-25.7	$7x10^{-3}$-$3x10^{-2}$	n.d.	Glaz et al. (2016)
Duckweed-based ponds						
USA	Synthetic stormwater	Temperate	1.7-3.3	0.5-1.9	$6.3x10^{-4}$-$4x10^{-2}$	Sims et al. (2013)
USA	Synthetic stormwater	Temperate	0.4-1.4	0.2-0.6	n.d.	Dai et al. (2015)

n.d. not data

1.6.2.2 Greenhouse gas emissions from AFPs

Five of the seven studies listed in **Table 1.4** provide data about the CH_4 flux in AFPs (Johansson *et al.*, 2004; Stadmark and Leonardson, 2005; Detweiler *et al.*, 2014; Hernandez-Paniagua *et al.*, 2014; Glaz *et al.*, 2016). As can be observed CH_4, CO_2 and N_2O production in AFPs is highly variable Although CO_2 assimilation by algal photosynthesis plays an important role in carbon dynamics, this gas was only measured and reported in two studies (Hernandez-Paniagua *et al.*, 2014; Glaz *et al.*, 2016). N_2O flux data was available for three studies (Johansson *et al.*, 2003; Singh *et al.*, 2005; Hernandez-Paniagua *et al.*, 2014).

Water temperature positively influences CO_2, CH_4, and N_2O fluxes from AFPs due to enhanced microbial activity (Singh *et al.*, 2005; Stadmark and Leonardson, 2005; Detweiler *et al.*, 2014). In studies into temperate conditions the highest emissions were observed during warmer summer months. In addition, substrate availability i.e. COD and NO_3^- also limits CH_4 and N_2O production (Johansson *et al.*, 2003; Johansson *et al.*, 2004; Stadmark and Leonardson, 2005). Nitrate concentrations between 8 and 16 mg NO_3-L^{-1} inhibits methane production (Stadmark and Leonardson, 2005; Stadmark and Leonardson, 2007). In AFPs a high uptake of nitrous oxide takes place when nitrate concentrations are low (<5 mg L^{-1}) (Johansson *et al.*, 2004). In shallow AFP, photosynthesis by algae increases oxygen concentrations in the water column, potentially inhibiting methanogenesis (Johansson *et al.*, 2004; Detweiler *et al.*, 2014); CH_4 oxidation efficiency for AFP system treating municipal wastewater has been estimated at 69.1% (Johansson *et al.*, 2004; Detweiler *et al.*, 2014).

1.6.2.3 Greenhouse gas emissions from duckweed based ponds

GHG flux data from DBPs containing municipal wastewater were hard to find in the literature. Although in a paper by Van der Steen *et al.* (2003) it was concluded that duckweed covers on stabilization ponds may reduce the emission of greenhouse gases, there were no data reported about GHG flux. The only two studies on CO_2 and CH_4 emissions were from two laboratory-scale duckweed ponds for storm water treatment (Sims *et al.*, 2013; Dai *et al.*, 2015). Although the simulated storm water had a low organic load, the CO_2 and CH_4 measured were higher than observed in some AFPs (**Table 1.4**). This behaviour was explained by low oxygen content in the water columns and the duckweed debris that acted as an additional organic source (Sims *et al.*, 2013). Likewise, when duckweed biomass was removed, CH_4 and CO_2 emissions decreased, suggesting that duckweed biomass served as a passive conduit for increased gas exchange between the soil, water and air interfaces (Sims *et al.*, 2013). This result is in contrast with findings reported by Van der Steen *et al.* (2003), who mentioned that a duckweed-cover reduced greenhouse gas emissions.

1.6.2.4 Greenhouse gas emissions from constructed wetlands

Constructed wetlands are natural-type systems for wastewater treatment designed to efficiently remove both organic matter and nutrients (Maucieri *et al.*, 2017). In literature, constructed wetlands are considered to be important sources of GHGs such as CO_2, CH_4, and N_2O. The studies reporting GHGs emissions in CWs involve domestic and dairy farm wastewater, and mining run-off (Tanner *et al.*, 1997; Fey *et al.*, 1999; Johansson *et al.*, 2003; Johansson *et al.*, 2004; Mander *et al.*, 2005; Teiter and Mander, 2005; Liikanen *et al.*, 2006; Sovik *et al.*, 2006; Gui *et al.*, 2007; Sovik and Klove, 2007; De Klein and Van der Werf, 2014; Wu *et al.*, 2016).

The studies reporting on CO_2 emissions in CWs are limited in number (Teiter and Mander, 2005; Liikanen *et al.*, 2006; Sovik *et al.*, 2006; Picek *et al.*, 2007; Ström *et al.* 2007; Van der Zaag *et al.*, 2010; Mander *et al.*, 2014). The values of CO_2 vary from -300 mg m^{-2} d^{-1} in a horizontal subsurface flow (HSSF) CW treating dairy farm wastewater (Van der Zaag *et al.*, 2010), up to 77,500 mg m^{-2} d^{-1} in a free water surface (FWS) CW treating municipal wastewater (Ström *et al.*, 2007). Different studies have suggested that CWs achieve significant C sequestration similar to natural wetlands (Mitsch et al., 2013). However, only a few studies reported CO_2 sequestration (Mander *et al.*, 2008; Van der Zaag *et al.*, 2010; Mander *et al.*, 2014), because the other studies had limitations, such as CO_2 measurements only from the soil or using opaque static chambers, in which CO_2 assimilation by photosynthesis cannot be assessed.

CH_4 is the most important greenhouse gas emitted from CWs. Methane fluxes from CWs ranged from -377 to 2,780 mg m^{-2} d^{-1} (**Table 1.5**). The negative flux for CH_4 reported in FWS CWs (Johansson *et al.*, 2004) suggests net consumption of this gas in this system. This consumption was observed in open areas where abundance of algae may lead to high oxygen levels and successively higher rates of bacterial methane oxidation (Johansson *et al.*, 2004). However, the high level of positive flux in most studies indicates that CWs are a significant source of CH_4 into the atmosphere.

Fluxes of N_2O have been measured in different types of CWs such as HSSF, VSSF, and FWS (**Table 1.5**). Nitrification and denitrification are the main pathways to produce N_2O emissions from CWs (Johansson *et al.*, 2003; Inamori *et al.*, 2007; Kampschreur *et al.*, 2009; Law *et al.*, 2012; Daelman *et al.*, 2013; Mander *et al.*, 2014). However, in CWs there are N_2O consumption, thus the highest consumption of N_2O (-8.4 mg m^{-2} d^{-1}) was observed in a FWS in Sweden (Johansson *et al.*, 2003), while the highest emission was reported in a HSSF in Norway (110 mg m^{-2} d^{-1}) (Sovik and Klove, 2007). N_2O consumption may be linked to a shortage of electron acceptors for denitrifying bacteria and is used as a substitute for nitrate (Maucieri *et al.*, 2017). This suggestion is indirectly confirmed by the fact that NO_3^- inhibits N_2O reduction in denitrification (Johansson *et al.*, 2003).

The GHG emissions from CWs are influenced by several factors including environmental conditions, CW type, wastewater characteristics and the capacity of the different plant species to assimilate nitrate and transfer oxygen (Maucieri *et al.*, 2017). Seasonal changes greatly affect the dynamics of CO_2, CH_4 and N_2O emitted from CWs due to changes in environmental conditions i.e. temperature and solar radiation (Liikanen *et al.*, 2006).

The emissions of CH_4, CO_2 and N_2O in HSSF and VSSF CWs were higher during summer than in other seasons (Sovik *et al.*, 2006; Van der Zaag *et al.*, 2010; Barbera *et al.*, 2014; Wu *et al.*, 2016). Stadmark and Leonardson (2005) described that water temperature below 15°C limited CH_4 flux from pond sediments. A similar conclusion was reported in an open field FWS CW with a threshold of 18°C limiting CH_4 and N_2O fluxes (Groh *et al.*, 2015). In a boreal FWS CW treating wastewater the CH_4 fluxes in summer were 10-50-fold higher than in other periods of the year due to the large temperature variations between seasons. In HSSF and VSSF CWs there was a significant correlation between water temperature and CO_2, CH_4 and N_2O emissions (Van der Zaag *et al.*, 2010; Mander *et al.*, 2014), although some other studies did not find a significant correlation between GHG flux and water temperature (Tanner *et al.*, 1997; Teiter and Mander, 2005; Picek *et al.*, 2007).

The type of wetland also affects GHG emission. VSSF CWs show slightly higher emissions of N_2O than HSSF (Teiter and Mander, 2005). However, CH_4 emission was higher in HSSF CWs than in VSSF, and non-significant differences were reported for CO_2 emissions (Maucieri *et al.*, 2017). The comparison of CO_2, CH_4, and N_2O fluxes in different CWs located in Northern Europe (Sovik *et al.*, 2006) showed that average N_2O emissions were significantly higher in the VSSF than HSSF and SSF CWs. During the summer, the CO_2 emission was significantly higher in VSSF than HSSF, and FWS. CWs treating dairy farm wastewater indicated that FWS CWs emitted significantly more CO_2, CH_4, and N_2O than HSSF CWs (Van der Zaag *et al.*, 2010).

These differences in emissions from different CWs can be attributed mainly to the different wastewater treatment conditions. For instance, the oxic conditions that predominate in VSSF CWs favour methane oxidation while the production of this gas is decreased. On the other hand, in FWS CWs anoxic conditions predominate, which reduces oxidation and increases methane production. In addition, the presence of oxygen in CWs leads to incomplete denitrification with higher N_2O emissions from VSSF than FWS CWs (Maucieri *et al.*, 2017). Organic carbon load positively influences the methane emissions in CWs. Higher gas emissions were measured under higher pollutant loading (Liikanen *et al.*, 2006; Chiemchaisri *et al.*, 2009; Van der Zaag *et al.*, 2010). Also Sovik and Klove (2007) reported that the flux of CH_4 in a HSSF CW was positively correlated with the influent total organic carbon (TOC_{in})

The carbon/nitrogen ratio (C/N) also influences GHG emissions in CWs. Synthetic wastewater with C/N ratio 5:1 provided a relatively high efficiency of organic matter removal and a low level of GHG flux in a VSSF CW (Yan *et al.*, 2012). Likewise, maximum nitrogen removal efficiency (NH_4^+-N 98% and TN 90%) and low N_2O emission (8.2 mg m^{-2} d^{-1}) accounting for 1.44 % of TN removal were obtained at C/N ratio 12:1 (Li *et al.*, 2017). Further, a COD/N ratio of 20 caused that N_2O emission to be ten times more than that measured under the COD/N ratio of 5 and 10 with 95.8% (Wu *et al.*, 2009).

Table 1.5 Methane, carbon dioxide, and nitrous oxide emissions in constructed wetlands for wastewater treatment. Average and range values for different sites and climate zones are presented. T-Temperate, B-Boreal, W-Winter

Type	Location	Wastewater	Climate	Plants	Measurement season	CO_2 $g.m^{-2}.d^-$	CH_4 $g.m^{-2}.d^{-1}$	N_2O $g.m^{-2}.d^{-1}$	References
HSSF	Estonia	Municipal	T/B	T.latifolia, P.australis	October-June	n.d.	-0.0002-2.8	0.0001-0.09	Mander (2003)
HSSF	Estonia	Municipal	T/B	T.latifolia, P.australis	Summer and winter	14±0.9[a]; 3.5±0.2[t]	0.5±0.32[a];2±0.4[b]	0.01±0.002[a]; 0.025±0.0004[b]	Teiter and Mander (2005)
HSSF	Norway	Municipal	B / B	No vegetation / Spagnum angustifolium	Summer and winter	2.9±0.6[a]; 1.0±0.2[t]	0.2±0.06[a];- 0.002±0.009	0.01±0.007[a]; 0.057±0.035[b]	Sovik et al. (2006)
HSSF	Finland	Mining runoff			All seasons (2001-2002)	0.09 – 33	0.0004-4.3	-0.001-0.003	Liikanen et al. (2006)
HSSF	Norway	Municipal	B	P.australis	Summer and winter	n.d.	-0.002 – 2.5	-0.0005-0.110	Sovik and Klove (2007)
HSSF	Czech	Municipal	T		All seasons	1.76-7.4	0-2.2	Negligible	Picek et al. (2007)
HSSF	Canada	Dairy farming	T	T. latifolia	August 2005-September 2006	-0.3-0.1	<1.024	<0.157	Vander Zaag et al. (2010)
VSSF	Estonia	Municipal	T/B	P.australis	Summer and winter	31±7.7[a]; 5.9±1.3[t]	0.15±0.0[a]; 0.045±0.008[b]	0.024±0.006[a]; 0.008±0.003[b]	Teiter and Mander (2005)
VSSF	Japan	Artificial WW	T	P.australis	All seasons	n.d.	0.480	0.001	Inamori et al. (2007)

Table 1.5 cont…

VSSF	Japan	Municipal	T	*P.australis*	All seasons	n.d.	0.1^c	0.003	Liu *et al.* (2009)
FWS	Sweden	Municipal	T	*Phalarys arundinacea, Lemna minor*	Spring and summer	n.d.	-0.4-1.7	-0.008-0.04	Johansson *et al.* (2003); Johansson *et al.* (2004)
FWS	Sweden	Municipal	T		2003 – 2004	n.d.	0.01-0.10	n.d.	Stadmark and Leonardson (2005)
FSW	Finland	Municipal	T	*Carex-Sphagnum*	Summer and winter	12.2 ± 1.4^a 4.8 ± 1.4^b	0.2 ± 0.01^a; $0.07\pm0.0b$	0.008 ± 0.002^a; 0.0003 ± 0.00005^b	Sovik *et al.* (2006)
FWS	Norway	Municipal	T	*T.latifolia*	Spring and summer	5.1 ± 0.9^a 3.6 ± 1.1^b	0.1 ± 0.04^a; 0.5 ± 0.3^b	0.006 ± 0.002^a; 0.002 ± 0.001^b	Sovik *et al* (2006)
FWS	Sweden	Municipal	T		Early summer	1.4-77.5	-0.4- 1.4	-0.016-0.03	Ström *et al.* (2007)
FWS	Sweden	Municipal	T		2003 – 2004	n.d.	0.01-1	n.d.	Stadmark and Leonardson (2005)

average value during summer; [b]: average value during winte

Nitrate concentration in wastewater can limit CH_4 and N_2O production in CWs. Nitrate concentrations of 8 and 16 $mg.L^{-1}$ can inhibit CH_4 production (Stadmark and Leonardson, 2005; (Stadmark and Leonardson, 2007). Groh *et al.* (2015) found a negative correlation between CH_4 flux and the nitrate concentration in a FWS CW that received tile drainage water. Although the correlation was relatively weak in the experiment was observed that when nitrate concentrations drop near to zero the highest methane concentration were observed (Maucieri *et al.*, 2017). On the other hand, NO_3^- inhibits N_2O reduction in denitrification process and there is even a greater atmospheric N_2O consumption (78%) when NO_3^- is below 0.5 $mg.L^{-1}$ (Johansson *et al.*, 2003).

The role of vegetation on GHG emissions from CWs has been discussed in different studies (Johansson *et al.*, 2004; Ström *et al.*, 2005; Inamori *et al.*, 2007; Wang *et al.*, 2013). Constructed wetlands may vary greatly in aquatic macrophytes species composition, comprising diverse typologies such as emergent and floating. Chang *et al.* (2014) reported that species richness increases nitrogen removal and N_2O emissions. In CWs planted with macrophytes such as *Typha latifolia* (Inamori *et al.*, 2007), *Z. latifolia* (Wang *et al.*, 2013), *Phalaris arundinacea* (Johansson *et al.*, 2004) and *Carex* (Ström *et al.*, 2005) it was observed that higher emissions occurred in vegetated areas. By contrast, the mat-cover formed by floating aquatic macrophytes represent a barrier against the escape of methane and carbon dioxide into the atmosphere in FWS (Johansson *et al.*, 2004; Stadmark and Leonardson, 2005; Mander *et al.*, 2014).

Plants may influence GHG emissions through the mechanisms of production, consumption and transport. Firstly, aquatic macrophytes release organic matter from root exudates and from the decaying litter, providing substrates for GHG production in the sediments (Liikanen *et al.*, 2006; Picek *et al.*, 2007; Ström *et al.*, 2007;Van der Zaag *et al.*, 2010). Increased species richness increases nitrogen removal and N_2O emissions, because species richness increases the biomass (organic carbon), which acts as an electron donor for nitrate under anaerobic conditions (Chang et al., 2014). Secondly, plants may reduce GHG produced in the sediments due to the oxidation of GHGs in the rhizosphere i.e. CH_4 (Johansson *et al.*, 2004). Thirdly, the plant aerenchyma may act as a conduit by which GHGs are conducted from the sediment to the atmosphere (Sovik *et al.*, 2006; Inamori *et al.*, 2007; Wang *et al.*, 2013; Mander *et al.*, 2014). Furthermore, the aerenchym cells form an important route for the transport of O_2 from the leaves to the roots and the O_2 may modify the soil oxidation-reduction status, which can increase CH_4 oxidation (Wang *et al.*, 2013).

1.7 QUANTIFYING EMISSIONS OF GREENHOUSE GASES FROM WASTEWATER

There is growing worldwide concern about global warming and climate change, because of their impacts on the environment and consequently on the people living on our planet. Excessive greenhouse gas emissions - mainly CO_2, CH_4 and N_2O - from anthropogenic sources are the main drivers for global warming and climate change (IPCC, 2013). That is why the identification and quantification of all sources, both natural and anthropogenic, is needed for developing strategies to control and reduce GHGs emissions into the atmosphere. Therefore, in a world with a growing awareness of the possible effects of human activities on climate change, the emission of greenhouse gases from wastewater and its treatment must be assessed for the following reasons:

- Wastewater contains organic compounds, which during degradation are converted into CO_2, CH_4, and N_2O. In wastewater treatment plants these conversions are accelerated, and they are considered as a GHG emission source. Wastewater treatment plants generate effluents containing organic carbon which has the potential to generate additional GHG emissions in the receiving water bodies, such as rivers and natural wetlands (Greenfield and Batstone, 2005). Depending on the type of WWT system used, aerobic, anoxic or combinations of both conditions may occur, affecting the production of CO_2, CH_4, and N_2O. Thus, wastewater management it is an important source of GHGs and must to be considered in the worldwide and national GHGs inventories.

- Factors such as population growth, urban intensification, and increasing coverage of water supply and sanitation in the future will also contribute to increased GHG emissions from the wastewater sector. These factors are increasing the amount of carbon flowing into municipal sewer networks and wastewater treatment systems, leading to increased CH_4, CO_2 and N_2O. However, the magnitude of these emissions depends of the treatment technology used. Most developing countries rely on less advanced wastewater treatment and collection systems such as latrines, septic tanks, open sewers, and lagoons, which may have higher GHG emissions, in particular CH_4. In addition, in developing countries only a limited percentage of the wastewater is treated and therefore untreated wastewater receive 'natural treatment' in the receiving water bodies, and will possibly generate even higher GHG emissions, due to the creation of anoxic conditions in the receiving water bodies.

- GHGs from wastewater have been estimated to account for 3-4 % of global emissions (Bogner et al., 2007; Ciais et al., 2014; Saunois et al., 2016). It has been estimated that between 1990 and 2025, global CH_4 and N_2O emissions from wastewater will have increased, from 352 to 477 $MtCO_{2eq}$ and from 82 to 100 $MtCO_{2eq}$, respectively (EPA, 2012; Karakurt et al., 2012) However, the contribution of the wastewater sector to GHG emissions may be underestimated, because most emissions are not measured directly, but are estimated through protocols that have been developed by the IPCC (2006). These protocols suggest calculating emissions by multiplying the metrics of activity in wastewater by the emission factors (EFs): the amount of GHG emitted per unit of activity. Although this methodology is quite simple, the EFs are an average of a broad range of conditions and often yield inaccurate estimates (Davidson and Kanter, 2014). There is a necessity for field data to estimate GHG emissions from wastewater and factors influencing their production.

The measurements in situ provide GHGs data with a lower uncertainty compare to emissions factor method, since under real conditions, the emissions of CO_2, CH_4, and N_2O in WWT are influenced by variables such as carbon substrate availability, dissolved oxygen concentration and the presence of potentially inhibitory intermediates (El-Fadel and Massoud, 2001).

- In developing countries there are many uncertainties with respect to direct emissions, indirect emissions and availability and quality of data from the wastewater sector (Bogner et al., 2007). Thus, it is necessary to develop research to estimate the emissions of CO_2, CH_4, and N_2O emitted from wastewater collection and treatment systems (Foley et al., 2015). This will allow to water authorities in these countries to address actions to control direct and indirect atmospheric GHGs from wastewater management sector (Bogner et al., 2007; Das, 2011).

1.8 SCOPE AND OBJECTIVES OF THIS THESIS

There is a global concern about the possible effects of human activities on global warming and therefore the identification of the GHGs from anthropogenic activities is a priority. In spite of increasing awareness of the significance of GHG emissions from the scientific community for this last decade, there are knowledge gaps mainly on data of GHGs emissions from tropical regions. There is also a lack of understanding of and CO_2, CH_4, and N_2O dynamics for EWWT operating under tropical conditions. Most studies were conducted under temperate conditions and only a few in tropical or subtropical regions, thus the production of CH_4, CO_2 and N_2O in EWWT operating under tropical conditions is largely unknown. The tropical conditions are characterized by high temperatures, long and stable photoperiods, photosynthetic activity high levels of bacterial and algal activity, and dynamics in dissolved oxygen and pH patterns which may all influence GHG dynamics differently compared to temperate conditions.

In order to develop a better understanding via scientific knowledge and comparable information on GHG emissions from EWWT systems under tropical conditions, the overall aim of this study was *to quantify and assess greenhouse gas emissions from different ecotechnologies for wastewater treatment in tropical countries*. The specific objectives of this research project were:

(1) To develop and to evaluate an analytical technique based on static chambers to estimate greenhouse gas emissions from EWWT operating under tropical conditions.

Under this specific objective the following research questions are addressed:

What is the comparability and reproducibility of GHG measurements using the closed static chamber technique?

What is the validity of linear regression to estimate GHG emissions from stabilization ponds using the static chamber technique by comparing the fluxes obtained from linear and non-linear models?

(2) To quantify GHGs from EWWT under tropical conditions.

This specific objective will address the following research question:

What type and quantity of GHGs are emitted from anaerobic ponds, algae facultative ponds, duckweed-based ponds and natural wetlands located in a tropical region i.e. Colombia?

(3) To assess the influence of operational parameters and environmental factors on the generation of GHGs from EWWT (anaerobic ponds, algae facultative ponds, duckweed-based ponds and natural wetlands).

This specific objective will address the following research questions:

Which are the environmental factors and operational conditions that control GHGs production in EWWT systems under tropical conditions?

What is the effect of environmental factors and operational conditions on GHG emissions produced in EWWT systems under tropical conditions?

1.9 DISSERTATION OUTLINE

Considering the specific objectives mentioned in Section 1.7, the thesis is divided into seven chapters.

Chapter 1, the general introduction provides a review of the climate system and the role of greenhouse gases, as well as the fundamental processes and scientific understandings behind GHG emissions from **EWWT**. **Chapter 2** assesses the static chamber technique to measure GHGs in facultative ponds. The validity of linear regression to estimate GHG emissions from stabilization ponds using the static chamber technique by comparing the fluxes obtained from linear and non-linear models is described. **Chapter 3** reports on CO_2, CH_4, and N_2O emissions in a full-scale AP; the influence of environmental and operational parameters such as pH, temperature and sludge accumulation in GHG production is discussed. The emissions of CO_2, CH_4, and N_2O from a pilot-scale AFP and a pilot-scale DBP are described in **Chapter 4**. Special emphasis was given to estimating differences in emissions during daytime (higher solar radiation) and night-time (lower solar radiation). In **Chapter 5** CO_2, CH_4, and N_2O emissions from a full-scale algal facultative pond (AFP) is described. The emissions of these gases were measured during daytime and night-time periods determining their daily variations. Besides, the influence of environmental parameters such as pH, DO, and temperature on GHG emissions was determined. In **Chapter 6**, the CO_2, CH_4, and N_2O emissions from a tropical eutrophic freshwater wetland receiving input loading from several sources i.e. agricultural run-off, domestic sewage, and a polluted river were studied. This study aims to look at similarities between engineered EWWT and natural systems and to confirm whether natural wetlands are influenced by anthropogenic activities and if they result in net emissions of greenhouse gases into the atmosphere. Since most wastewaters in developing countries are discharged without treatment, it is necessary to estimate the impact on GHG emissions in receiving water bodies. **Chapter 7** presents a discussion of the main results, which is followed by conclusions and recommendations.

1.10 REFERENCES

Ahn, J.H., Kim, S., Park, H., Rahm, B., Pagilla, K. and Chandran, K. (2010). N_2O emissions from activated sludge processes, 2008– 2009: results of a national monitoring survey in the United States. Environmental Science & Technology 44(12), 4505-4511.

Alaerts, G.J., Mahbubar, R. and Kelderman, P. (1996). Performance analysis of a full-scale duckweed-covered sewage lagoon. Water research 30, 843-852.

Arthur, J.P. (1983) Notes on the design and operation of waste stabilization ponds in warm climates of developing countries. Technical Paper 7, The World Bank, Washington, DC.

Barbera, A.C., Borin, M., Ioppolo, A., Cirelli, G.L. and Maucieri, C. (2014). Carbon dioxide emissions from horizontal sub-surface constructed wetlands in the Mediterranean Basin. Ecological Engineering 64, 57-61.

Barton, P.K. and Atwater, J.W. (2002). Nitrous oxide emissions and the anthropogenic nitrogen in wastewater and solid waste. Journal of Environmental Engineering 128, 137-150.

Bogner, J., Abderalfie, A., Diaz, C., Faaij, A., Gao, Q., Hashimoto, S., Mareckova, K., Pipatti, R. and Zhang, T. (2007). Waste Management, In Climate Change 2007: Mitigation. Contribution of Working Group III to the Fourth Assessment Report of the Intergovernmental Panel on Climate Change. B. Metz, O.R. Davidson, P.R. Bosch, R. Dave and Meyer, L.A. (eds), p. 32, Cambridge University, Cambridge, United Kingdom

Bogner, J., Pipatti, R., Hashimoto, S., Diaz, C., Mareckova, K., Diaz, L., Kjeldsen, P., Monni, S., Faaij, A. and Gao, Q. (2008). Mitigation of global greenhouse gas emissions from waste: conclusions and strategies from the Intergovernmental Panel on Climate Change (IPCC) Fourth Assessment Report. Working Group III (Mitigation). Waste Management & Research 26(1), 11-32.

Caicedo, J.R. (2005). Effect of Operational Variables on Nitrogen Transformations in Duckweed Stabilization Ponds. PhD, UNESCO-IHE- Wageningen University, Delft.

Campos, J., Valenzuela-Heredia, D., Pedrouso, A., Val del Río, A., Belmonte, M. and Mosquera-Corral, A. (2016). Greenhouse Gases Emissions from Wastewater Treatment Plants: Minimization, Treatment, and Prevention. Journal of Chemistry 2016, 1-12.

Chang, J., Fan, X., Sun, H., Zhang, C., Song, C., Chang, S.X., Gu, B., Liu, Y., Li, D. and Wang, Y. (2014). Plant species richness enhances nitrous oxide emissions in microcosms of constructed wetlands. Ecological Engineering 64, 108-115.

Chiemchaisri, C., Chiemchaisri, W., Junsod, J., Threedeach, S. and Wicranarachchi, P.N. (2009). Leachate treatment and greenhouse gas emission in subsurface horizontal flow constructed wetland. Bioresource Technology 100, 3808-3814.

Ciais, P., Sabine, C., Bala, G., Bopp, L., Brovkin, V., Canadell, J., Chhabra, A., DeFries, R., Galloway, J. and Heimann, M. (2014). Climate change 2013: the physical science basis. Contribution of Working Group I to the Fifth Assessment Report of the Intergovernmental Panel on Climate Change, pp. 465-570, Cambridge University Press.

Craggs, R., Park, J. and Heubeck, S. (2008). Methane emissions from anaerobic ponds on a piggery and a dairy farm in New Zealand. Animal Production Science 48, 142-146.

Czepiel, P., Crill, P.M. and Harriss, R.C. (1995). Nitrous oxide emissions from municipal wastewater treatment. Environmental Science & Technology 29(9), 2352-2356.

Czepiel, P.M., Crill, P.M. and Harriss, R.C. (1993). Methane emissions from municipal wastewater treatment processes. Environmental Science & Technology 27(12), 2472-2477.

Daelman, M.R., van Voorthuizen, E.M., van Dongen, U.G., Volcke, E.I. and van Loosdrecht, M.C. (2012). Methane emission during municipal wastewater treatment. Water research 46(11), 3657-3670.

Daelman, M.R.J., van Voorthuizen, E.M., van Dongen, L., Volcke, E.I.P. and van Loosdrecht, M.C.M. (2013). Methane and nitrous oxide emissions from municipal wastewater treatment - results from a long-term study. Water Science and Technology 67, 2350-2355.

Dai, J., Zhang, C., Lin, C.-H. and Hu, Z. (2015). Emission of Carbon Dioxide and Methane from Duckweed Ponds for Stormwater Treatment. Water Environment Research 87, 805-812.

Das, S. (2011). Estimation of Greenhouse Gases Emissions from Biological Wastewater Treatment Plants at Windsor. Electronic Theses and Dissertations. 77. https://scholar.uwindsor.ca/etd/77

Davidson, E.A. and Kanter, D. (2014). Inventories and scenarios of nitrous oxide emissions. Environmental Research Letters 9, 105012.

De Klein, J.J. and Van der Werf, A.K. (2014). Balancing carbon sequestration and GHG emissions in a constructed wetland. Ecological Engineering 66, 36-42.

De Sutter, T.M. and Ham, J.M. (2005). Lagoon-biogas emissions and carbon balance estimates of a swine production facility. Journal of environmental quality 34(1), 198-206.

Detweiler, A.M., Bebout, B.M., Frisbee, A.E., Kelley, C.A., Chanton, J.P. and Prufert-Bebout, L.E. (2014). Characterization of methane flux from photosynthetic oxidation ponds in a wastewater treatment plant. Water Science & Technology 70, 980-989.

Dlugokencky, E.J., Bruhwiler, L., White, J.W.C., Emmons, L.K., Novelli, P.C., Montzka, S.A., Masarie, K.A., Lang, P.M., Crotwell, A.M., Miller, J.B. and Gatti, L.V. (2009). Observational constraints on recent increases in the atmospheric CH_4 burden. Geophysical Research Letters 36.

EDGAR-database (2009). Emission Database for Global Atmospheric Research (EDGAR), release version 4.0. European Commission. Joint Research Centre (JRC) / Netherlands Environmental Assessment Agency (PBL).

El-Fadel, M. and Massoud, M. (2001). Methane emissions from wastewater management. Environmental pollution 114, 177-185.

El-Shafai, S.A., El-Gohary, F.A., Nasr, F.A., Peter van der Steen, N. and Gijzen, H.J. (2007). Nutrient recovery from domestic wastewater using a UASB-duckweed ponds system. Bioresource Technology 98, 798-807.

EPA (2010). Methane and nitrous oxide emissions from natural sources. United States. Environmental Protection Agency (EPA) Report, Washington, D.C.

EPA (2012). Global Anthropogenic Non-CO2 Greenhouse Gas Emissions: 1990–2030. Office of Atmospheric Programs Climate Change Division, Washington D.C: US Environmental Protection Agency.US EPA Washington, DC.

EPA (2016). Inventory of U.S. greenhouse gas emissions and sinks: 1990-2012, Environmental Protection Agency EPA, Washington.

Fey, A., Benckiser, G. and Ottow, J.C.G. (1999). Emissions of nitrous oxide from a constructed wetland using a groundfilter and macrophytes in waste-water purification of a dairy farm. Biology and fertility of soils 29, 354-359.

Foley, J., Yuan, Z., Keller, J., Senante, E., Chandran, K., Willis, J., Shah, A., van Loosdrecht, M. and van Voorthuizen, E. (2011). N_2O and CH_4 emission from wastewater collection and treatment systems: technical report.

Foley, J., Yuan, Z., Keller, J., Senante, E., Chandran, K., Willis, J., Shah, A., van Loosdrecht, M.C. and van Voorthuizen, E. (2015) N_2O and CH_4 Emission from Wastewater Collection and Treatment Systems: State of the Science Report and Technical Report, IWA Publishing.

Forster, P., V. Ramaswamy, P. Artaxo, T. Berntsen, R. Betts, D.W. Fahey, J. Haywood, J. Lean, D.C. Lowe, G. Myhre, J. Nganga, R. Prinn, G. Raga, M. Schulz and R. Van Dorland, (2007). Changes in Atmospheric Constituents and in Radiative Forcing In: Climate Change 2007: The Physical Science Basis. Contribution of Working Group I to the Fourth Assessment Report of the Intergovernmental Panel on Climate Change [Solomon, S., D. Qin, M. Manning, Z. Chen, M. Marquis, K.B. Averyt, M.Tignor and H.L. Miller (eds.)]. Cambridge University Press, Cambridge, United Kingdom and New York, NY, USA., IPCC, Cambridge, United Kingdom and New York, NY, USA.

Ghosh, A., Patra, P.K., Ishijima, K., Umezawa, T., Ito, A., Etheridge, D.M., Sugawara, S., Kawamura, K., Miller, J.B., Dlugokencky, E.J., Krummel, P.B., Fraser, P.J., Steele, L.P., Langenfelds, R.L., Trudinger, C.M., White, J.W.C., Vaughn, B., Saeki, T., Aoki, S. and Nakazawa, T. (2015). Variations in global methane sources and sinks during 1910-2010. Atmospheric Chemistry and Physics 15, 2595-2612.

Gijzen, H. (2001). Anaerobes, aerobes and phototrophs - A winning team for wastewater management. Water Science and Technology 44(8), 123-132.

Gijzen, H. (2002). Anaerobic digestion for sustainable development: a natural approach. Water Science and Technology 45(10), 321-328.

Glaz, P., Bartosiewicz, M., Laurion, I., Reichwaldt, E.S., Maranger, R. and Ghadouani, A. (2016). Greenhouse gas emissions from waste stabilisation ponds in Western Australia and Quebec (Canada). Water research 101, 64-74.

Goldberg, S.D. and Gebauer, G. (2009). Drought turns a Central European Norway spruce forest soil from an N_2O source to a transient N_2O sink. Global Change Biology 15(4), 850-860.

Greenfield, P. and Batstone, D.J. (2005). Anaerobic digestion: impact of future greenhouse gases mitigation policies on methane generation and usage. Water Science and Technology 52, 39-47.

Groh, T.A., Gentry, L.E. and David, M.B. (2015). Nitrogen removal and greenhouse gas emissions from constructed wetlands receiving tile drainage water. Journal of environmental quality 44, 1001-1010.

Gui, P., Inamori, R., Matsumura, M. and Inamori, Y. (2007). Evaluation of constructed wetlands by wastewater purification ability and greenhouse gas emissions. Water Science & Technology 56, 49-55.

Gujer, W. and Zehnder, A.J. (1983). Conversion processes in anaerobic digestion. Water Science and Technology 15(8-9), 127-167.

Guo, G., Wang, Y., Hao, T., Wu, D. and Chen, G.-H. (2018). Enzymatic nitrous oxide emissions from wastewater treatment. Frontiers of Environmental Science & Engineering 12(1), 10.

Hartmann, D.L., Tank, A.M.K., Rusticucci, M., Alexander, L.V., Brönnimann, S., Charabi, Y.A.R., Dentener, F.J., Dlugokencky, E.J., Easterling, D.R. and Kaplan, A. (2013). Climate Change 2013 the Physical Science Basis: Working Group I Contribution to the Fifth Assessment Report of the Intergovernmental Panel on Climate Change, Cambridge University Press.

Hasanudin, U., Utomo, T., Suroso, E., Hendri, J., Inokawa, A. and Fujie, K. (2006). Methane and CO_2 Gases Emission from Anaerobic Pond in Tapioca Wastewater Treatment, in 7th IWA Specialist Conference on Waste Stabilization Ponds, Bangkok, Thailand.

Hernandez-Paniagua, I.Y., Ramirez-Vargas, R., Ramos-Gomez, M.S., Dendooven, L., Avelar-Gonzalez, F.J. and Thalasso, F. (2014). Greenhouse gas emissions from stabilization ponds in subtropical climate. Environmental Technology 35, 727-734.

Ho, L.T., Van Echelpoel, W. and Goethals, P.L. (2017). Design of waste stabilization pond systems: A review. Water research.

Hwang, K.-L., Bang, C.-H. and Zoh, K.-D. (2016). Characteristics of methane and nitrous oxide emissions from the wastewater treatment plant. Bioresource Technology 214, 881-884.

Inamori, R., Gui, P., Dass, P., Matsumura, M., Xu, K.-Q., Kondo, T., Ebie, Y. and Inamori, Y. (2007). Investigating CH 4 and N 2 O emissions from eco-engineering wastewater treatment processes using constructed wetland microcosms. Process Biochemistry 42(3), 363-373.

IPCC (2006). 2006 IPCC guidelines for national greenhouse gas inventories. Eggleston H.S., Buendia, L., Miwa K., Ngara T. and Tanabe K. (ed), Intergovernmental Panel on Climate Change, Hayama, Japan.

IPCC (2013). Summary for Policymakers. In: Climate Change 2013: The Physical Science Basis. Contribution of Working Group I to the Fifth Assessment Report of the Intergovernmental Panel on Climate Change [Stocker, T.F., D. Qin, G.-K. Plattner, M. Tignor, S.K. Allen, J. Boschung, A. Nauels, Y. Xia, V. Bex and P.M. Midgley (eds.)]. Cambridge University Press,, Cambridge, United Kingdom and New York, NY, USA.

IPCC (2014). Climate Change 2014: Synthesis Report. Contribution of Working Groups I, II and III to the Fifth Assessment Report of the Intergovernmental Panel on Climate Change [Core Writing Team, R.K. Pachauri and L.A. Meyer (eds.)]. IPCC, Geneva, Switzerland, 151 pp.

Johansson, A., Gustavsson, A.-M., Öquist, M. and Svensson, B. (2004). Methane emissions from a constructed wetland treating wastewater—seasonal and spatial distribution and dependence on edaphic factors. Water research 38(18), 3960-3970.

Johansson, A., Klemedtsson, Å.K., Klemedtsson, L. and Svensson, B. (2003). Nitrous oxide exchanges with the atmosphere of a constructed wetland treating wastewater. Tellus B 55(3), 737-750.

Joss, A., Salzgeber, D., Eugster, J., König, R., Rottermann, K., Burger, S., Fabijan, P., Leumann, S., Mohn, J. and Siegrist, H. (2009). Full-scale nitrogen removal from digester liquid with partial nitritation and anammox in one SBR. Environmental Science & Technology 43(14), 5301-5306.

Kadlec, R.H., Tanner, C.C., Hally, V.M. and Gibbs, M.M. (2005). Nitrogen spiraling in subsurface-flow constructed wetlands: Implications for treatment response. Ecological Engineering 25(4), 365-381.

Kampschreur, M.J., Temmink, H., Kleerebezem, R., Jetten, M.S. and van Loosdrecht, M.C. (2009). Nitrous oxide emission during wastewater treatment. Water research 43(17), 4093-4103.

Karakurt, I., Aydin, G. and Aydiner, K. (2012). Sources and mitigation of methane emissions by sectors: A critical review. Renewable energy 39(1), 40-48.

Konaté, Y., Maiga, A.H., Casellas, C. and Picot, B. (2013). Biogas production from an anaerobic pond treating domestic wastewater in Burkina Faso. Desalination and Water Treatment 51(10-12), 2445-2452.

Kroeze, C., Mosier, A. and Bouwman, L. (1999). Closing the global N_2O budget: a retrospective analysis 1500–1994. Global Biogeochemical Cycles 13(1), 1-8.

Law, Y., Ye, L., Pan, Y. and Yuan, Z. (2012). Nitrous oxide emissions from wastewater treatment processes. Phil. Trans. R. Soc. B 367(1593), 1265-1277.

Lens, P., Zeeman, G. and Lettinga, G. (2001) Decentralised sanitation and reuse, IWA publishing.

Li, M., Wu, H., Zhang, J., Ngo, H.H., Guo, W. and Kong, Q. (2017). Nitrogen removal and nitrous oxide emission in surface flow constructed wetlands for treating sewage treatment plant effluent: Effect of C/N ratios. Bioresource Technology.

Liikanen, A., Huttunen, J.T., Karjalainen, S.M., Heikkinen, K., Vaisanen, T.S., Nykanen, H. and Martikainen, P.J. (2006). Temporal and seasonal changes in greenhouse gas emissions from a constructed wetland purifying peat mining runoff waters. Ecological Engineering 26(3), 241-251.

Liu, C., Xu, K., Inamori, R., Ebie, Y., Liao, J. and Inamori, Y. (2009). Pilot-scale studies of domestic wastewater treatment by typical constructed wetlands and their greenhouse gas emissions. Frontiers of Environmental Science & Engineering in China 3(4), 477-482.

Mander, Ü., Dotro, G., Ebie, Y., Towprayoon, S., Chiemchaisri, C., Nogueira, S.F., Jamsranjav, B., Kasak, K., Truu, J. and Tournebize, J. (2014). Greenhouse gas emission in constructed wetlands for wastewater treatment: a review. Ecological Engineering 66, 19-35.

Mander, Ü., Lõhmus, K., Teiter, S., Mauring, T., Nurk, K. and Augustin, J. (2008). Gaseous fluxes in the nitrogen and carbon budgets of subsurface flow constructed wetlands. Science of The Total Environment 404(2), 343-353.

Mander, U., Teiter, S. and Augustin, J. (2005). Emission of greenhouse gases from constructed wetlands for wastewater treatment and from riparian buffer zones. Water Sci Technol 52(10-11), 167-176.

Mara, D. (2005). Pond treatment technology. Shilton, A. (ed), pp. 168-187, IWA Publishing London.

Maucieri, C., Barbera, A.C., Vymazal, J. and Borin, M. (2017). A review on the main affecting factors of greenhouse gases emission in constructed wetlands. Agricultural and Forest Meteorology 236, 175-193.

Mitsch, W.J., Bernal, B., Nahlik, A.M., Mander, Ü., Zhang, L., Anderson, C.J., Jørgensen, S.E. and Brix, H. (2013). Wetlands, carbon, and climate change. Landscape Ecology, 1-15.

Montzka, S., Dlugokencky, E. and Butler, J. (2011). Non-CO_2 greenhouse gases and climate change. Nature 476 (7358), 43-50.

Mosier, A. (1998). Soil processes and global change. Biology and fertility of soils 27(3), 221-229.

Myhre, G., Shindell, D., Bréon, F.-M., Collins, W., Fuglestvedt, J., Huang, J., Koch, D., Lamarque, J.-F., Lee, D. and Mendoza, B. (2013). Anthropogenic and natural radiative forcing. Climate change 423.

Nhapi, I., Dalu, J., Ndamba, J., Siebel, M.A. and Gijzen, H.J. (2003). An evaluation of duckweed-based pond systems as an alternative option for decentralised treatment and reuse of wastewater in Zimbabwe. Water Sci Technol 48(2), 323-330.

Nhapi, I. and Gijzen, H.J. (2005). A 3-step strategic approach to sustainable wastewater management. Water SA 31(1), 133-140.

Nisbet, E.G., Dlugokencky, E.J. and Bousquet, P. (2014). Methane on the Rise-Again. Science 343(6170), 493-495.

Paredes, M.G., Güereca, L.P., Molina, L.T. and Noyola, A. (2015). Methane emissions from stabilization ponds for municipal wastewater treatment in Mexico. Journal of Integrative Environmental Sciences, 1-15.

Paustian, L., Babcock, B., Hatfield, J.L., Lal, R., McCarl, B.A., McLaughlin, S., Mosier, A., Rice, C., Roberton, G. and Rosenberg, N. (2016). Agricultural mitigation of greenhouse gases: science and policy options, Washington, DC: Conference on Carbon Sequestration.

Pearson, H., Mara, D., Cawley, L., Arridge, H. and Silva, S. (1996). The performance of an innovative tropical experimental waste stabilisation pond system operating at high organic loadings. Water Science and Technology 33(7), 63-73.

Peña, M.R., Madera, C.A. and Mara, D.D. (2002). Feasibility of waste stabilization pond technology for small municipalities in Colombia. Water Science & Technology 45(1), 1-8.

Picek, T., Čížková, H. and Dušek, J. (2007). Greenhouse gas emissions from a constructed wetland—Plants as important sources of carbon. Ecological Engineering 31(2), 98-106.

Picot, B., Paing, J., Sambuco, J.P., Costa, R.H.R. and Rambaud, A. (2003). Biogas production, sludge accumulation and mass balance of carbon in anaerobic ponds. Water Science and Technology 48(2), 243-250.

Prather, M.J., Holmes, C.D. and Hsu, J. (2012). Reactive greenhouse gas scenarios: Systematic exploration of uncertainties and the role of atmospheric chemistry. Geophysical Research Letters 39(9).

Ravindranath, N.H. and Sathaye, J.A. (2002). Climate Change and Developing Countries, pp. 247-265, Springer.

Ravishankara, A., Daniel, J.S. and Portmann, R.W. (2009). Nitrous oxide (N_2O): the dominant ozone-depleting substance emitted in the 21st century. Science 326(5949), 123-125.

Rousseau, D.P.L., Vanrolleghem, P.A. and Pauw, N. (2004). Constructed wetlands in Flanders: a performance analysis. Ecological Engineering 23(3), 151-163.

Saunois, M., Bousquet, P., Poulter, B., Peregon, A., Ciais, P., Canadell, J.G., Dlugokencky, E.J., Etiope, G., Bastviken, D., Houweling, S., Janssens-Maenhout, G., Tubiello, F.N., Castaldi, S., Jackson, R.B., Alexe, M., Arora, V.K., Beerling, D.J., Bergamaschi, P., Blake, D.R., Brailsford, G., Brovkin, V., Bruhwiler, L., Crevoisier, C., Crill, P., Covey, K., Curry, C., Frankenberg, C., Gedney, N., Hoglund-Isaksson, L., Ishizawa, M., Ito, A., Joos, F., Kim, H.S., Kleinen, T., Krummel, P., Lamarque, J.F., Langenfelds, R., Locatelli, R., Machida, T., Maksyutov, S., McDonald, K.C., Marshall, J., Melton, J.R.,

Morino, I., Naik, V., O'Doherty, S., Parmentier, F.J.W., Patra, P.K., Peng, C.H., Peng, S.S., Peters, G.P., Pison, I., Prigent, C., Prinn, R., Ramonet, M., Riley, W.J., Saito, M., Santini, M., Schroeder, R., Simpson, I.J., Spahni, R., Steele, P., Takizawa, A., Thornton, B.F., Tian, H.Q., Tohjima, Y., Viovy, N., Voulgarakis, A., van Weele, M., van der Werf, G.R., Weiss, R., Wiedinmyer, C., Wilton, D.J., Wiltshire, A., Worthy, D., Wunch, D., Xu, X.Y., Yoshida, Y., Zhang, B., Zhang, Z. and Zhu, Q. (2016). The global methane budget 2000-2012. Earth System Science Data 8(2), 697-751.

Sekomo, C.B., Rousseau, D.P., Saleh, S.A. and Lens, P.N. (2012). Heavy metal removal in duckweed and algae ponds as a polishing step for textile wastewater treatment. Ecological Engineering 44, 102-110.

Shilton, A. and Walmsey, N. (2005). Pond treatment technology. Shilton, A. (ed), pp. 1-13, IWA Publishing, London.

Sims, A., Gajaraj, S. and Hu, Z. (2013). Nutrient removal and greenhouse gas emissions in duckweed treatment ponds. Water research 47(3), 1390-1398.

Singh, V.P., Dass, P., Kaur, K., Billore, S.K., Gupta, P.K. and Parashar, D.C. (2005). Nitrous oxide fluxes in a tropical shallow urban pond under influencing factors. Current Science 88(3), 478.

Sovik, A.K., Augustin, J., Heikkinen, K., Huttunen, J.T., Necki, J.M., Karjalainen, S.M., Klove, B., Liikanen, A., Mander, U. and Puustinen, M. (2006). Emission of the Greenhouse Gases Nitrous Oxide and Methane from Constructed Wetlands in Europe. Journal of environmental quality 35(6), 2360.

Sovik, A.K. and Klove, B. (2007). Emission of N_2O and CH_4 from a constructed wetland in southeastern Norway. Sci Total Environ.

Spahni, R., Wania, R., Neef, L., Weele, M.v., Pison, I., Bousquet, P., Frankenberg, C., Foster, P., Joos, F. and Prentice, I. (2011). Constraining global methane emissions and uptake by ecosystems. Biogeosciences 8(6), 1643-1665.

Stadmark, J. and Leonardson, L. (2005). Emissions of greenhouse gases from ponds constructed for nitrogen removal. Ecological Engineering 25(5), 542-551.

Stadmark, J. and Leonardson, L. (2007). Greenhouse gas production in a pond sediment: Effects of temperature, nitrate, acetate and season. Science of the Total Environment 387(1-3), 194-205.

Ström, L., Lamppa, A. and Christensen, T.R. (2007). Greenhouse gas emissions from a constructed wetland in southern Sweden. Wetlands Ecology and Management 15(1), 43-50.

Ström, L., Mastepanov, M. and Christensen, T.R. (2005). Species-specific effects of vascular plants on carbon turnover and methane emissions from wetlands. Biogeochemistry 75(1), 65-82.

Tallec, G., Garnier, J., Billen, G. and Gousailles, M. (2008). Nitrous oxide emissions from denitrifying activated sludge of urban wastewater treatment plants, under anoxia and low oxygenation. Bioresource Technology 99(7), 2200-2209.

Tanner, C.C., Adams, D.D. and Downes, M.T. (1997). Methane emissions from constructed wetlands treating agricultural wastewaters. Journal of environmental quality 26(4), 1056-1062.

Tarasova, O., Koide, H. and Dlugokencky, E. (2016). The state of greenhouse gases in the atmosphere using global observations through 2014.EGU General Assembly Conference Abstracts. Geophysical Research Abstracts 18, 14837.

Teiter, S. and Mander, Ü. (2005). Emission of N_2O, N_2, CH_4, and CO_2 from constructed wetlands for wastewater treatment and from riparian buffer zones. Ecological Engineering 25(5), 529-542.

Thomson, A.J., Giannopoulos, G., Pretty, J., Baggs, E.M. and Richardson, D.J. (2012). Biological sources and sinks of nitrous oxide and strategies to mitigate emissions. Philosophical Transactions of the Royal Society B: Biological Sciences 367(1593), 1157-1168.

Toprak, H.k. (1995). Temperature and organic loading dependency of methane and carbon dioxide emission rates of a full-scale anaerobic waste stabilization pond. Water research 29(4), 1111-1119.

UNEP (2013). Drawing Down N_2O to Protect Climate and the Ozone Layer. A UNEP Synthesis Report. United Nations Environment Programme (UNEP), Nairobi, Kenya. http://www.unep.org/publications/ebooks/UNEPN2Oreport/

Van der Steen, N.P., Nakiboneka, P., Mangalika, L., Ferrer, A.V. and Gijzen, H.J. (2003). Effect of duckweed cover on greenhouse gas emissions and odour release from waste stabilisation ponds. Water Sci Technol 48(2), 341-348.

Van der Zaag, A.C., Gordon, R.J., Burton, D.L., Jamieson, R.C. and Stratton, G.W. (2010). Greenhouse gas emissions from surface flow and subsurface flow constructed wetlands treating dairy wastewater. Journal of environmental quality 39(2), 460-471.

Verbyla, M., Iriarte, M., Guzmán, A.M., Coronado, O., Almanza, M. and Mihelcic, J. (2016). Pathogens and fecal indicators in waste stabilization pond systems with direct reuse for irrigation: Fate and transport in water, soil and crops. Science of The Total Environment 551, 429-437.

Verma, R. and Suthar, S. (2016). Bioenergy potential of duckweed (Lemna gibba L.) biomass. Energy Sources, Part A: Recovery, Utilization, and Environmental Effects 38(15), 2231-2237.

Verstraete, W. and Vlaeminck, S.E. (2011). ZeroWasteWater: short-cycling of wastewater resources for sustainable cities of the future. International Journal of Sustainable Development & World Ecology 18(3), 253-264.

Von Sperling, M. and Chernicharo, C.A. (2005) Biological wastewater treatment in warn climate regions, IWA Publishing, London.

Vymazal, J. (2007). Removal of nutrients in various types of constructed wetlands. Science of The Total Environment 380(1), 48-65.

Vymazal, J. and Březinová, T. (2016). Accumulation of heavy metals in aboveground biomass of Phragmites australis in horizontal flow constructed wetlands for wastewater treatment: A review. Chemical Engineering Journal 290, 232-242.

Vymazal, J. and Kröpfelová, L. (2009). Removal of organics in constructed wetlands with horizontal sub-surface flow: a review of the field experience. Science of The Total Environment 407(13), 3911-3922.

Wang, J., Zhang, J., Xie, H., Qi, P., Ren, Y. and Hu, Z. (2011). Methane emissions from a full-scale A/A/O wastewater treatment plant. Bioresource Technology 102(9), 5479-5485.

Wang, Y., Yang, H., Ye, C., Chen, X., Xie, B., Huang, C., Zhang, J. and Xu, M. (2013). Effects of plant species on soil microbial processes and CH_4 emission from constructed wetlands. Environmental pollution 174, 273-278.

Wu, H., Lin, L., Zhang, J., Guo, W., Liang, S. and Liu, H. (2016). Purification ability and carbon dioxide flux from surface flow constructed wetlands treating sewage treatment plant effluent. Bioresource Technology 219, 768-772.

Wu, J., Zhang, J., Jia, W., Xie, H., Gu, R.R., Li, C. and Gao, B. (2009). Impact of COD/N ratio on nitrous oxide emission from microcosm wetlands and their performance in removing nitrogen from wastewater. Bioresource Technology 100(12), 2910-2917.

Wunderlin, P., Mohn, J., Joss, A., Emmenegger, L. and Siegrist, H. (2012). Mechanisms of N_2O production in biological wastewater treatment under nitrifying and denitrifying conditions. Water research 46(4), 1027-1037.

Yacob, S., Hassan, M.A., Shirai, Y., Wakisaka, M. and Subash, S. (2005). Baseline study of methane emission from open digesting tanks of palm oil mill effluent treatment. Chemosphere 59(11), 1575-1581.

Yan, C., Zhang, H., Li, B., Wang, D., Zhao, Y. and Zheng, Z. (2012). Effects of influent C/N ratios on CO_2 and CH_4 emissions from vertical subsurface flow constructed wetlands treating synthetic municipal wastewater. Journal of hazardous materials 203, 188-194.

Young, P., Archibald, A., Bowman, K., Lamarque, J.-F., Naik, V., Stevenson, D., Tilmes, S., Voulgarakis, A., Wild, O. and Bergmann, D. (2013). Pre-industrial to end 21st century projections of tropospheric ozone from the Atmospheric Chemistry and Climate Model Intercomparison Project (ACCMIP). Atmospheric Chemistry and Physics 13(4), 2063-2090.

Yusuf, R.O., Noor, Z.Z., Abba, A.H., Hassan, M.A.A. and Din, M.F.M. (2012). Methane emission by sectors: a comprehensive review of emission sources and mitigation methods. Renewable and Sustainable Energy Reviews 16(7), 5059-5070.

Zimmo, O., Van Der Steen, N. and Gijzen, H.J. (2003). Comparison of ammonia volatilisation rates in algae and duckweed-based waste stabilisation ponds treating domestic wastewater. Water research 37(19), 4587-4594.

Chapter 2

Flux estimation of greenhouse gases from wastewater stabilization ponds using the static chamber technique: Comparison of linear and non-linear models

This Chapter has been presented and published as:

Silva J.P., Lasso AP, Lubberding HJ, Peña MR, and Gijzen HJ (2015). **Biases in Greenhouse Gases Static Chambers Measurements in Stabilization Ponds: Comparison of Flux Estimation Using Linear and non-Linear Models.** *Atmospheric Environment* Vol 109, 130-138.

Silva J.P., Lasso AP, Lubberding HJ, Peña MR, and Gijzen HJ (2009). **Adaptation of a methodology for sampling GHGE from sewage treatment systems in Colombia** in: Air and Waste Management Conference "International Greenhouse Gas Measurement Symposium", March 23-25, 2009, Burligane, CA.

Silva J.P., Lasso AP, Lubberding HJ, Peña MR, Lens PNL and Gijzen HJ (2009). **Estimation of Greenhouse Gas Emissions by Static Chambers in Stabilization Ponds: The mixing Effect** In: 9th IWA stabilization ponds, August 1-3, 2011, Adelaide, AU.

Abstract

The closed static chamber technique is widely used to quantify greenhouse gases (GHG) i.e. CH_4, CO_2, and N_2O from aquatic and wastewater treatment systems. However, chamber-measured fluxes over air-water interfaces appear to be subject to considerable uncertainty, depending on the chamber design, lack of air mixing in the chamber, concentration gradient changes during the deployment, and irregular eruptions of gas accumulated in the sediment. In this study, the closed static chamber technique was tested in an anaerobic pond operating under tropical conditions. The closed static chambers were found to be reliable to measure GHG, but an intrinsic limitation of using closed static chambers is that not all the data for gas concentrations measured within a chamber headspace can be used to estimate the flux due to gradient concentration curves with non-plausible and physical explanations. Based on the total data set (n = 47), the percentage of curves accepted were 93.6, 87.2, and 73% for CH_4, CO_2 and N_2O, respectively. The statistical analyses demonstrated that only considering linear regression was frequently inappropriate for the determination of GHG flux from stabilization ponds by the closed static chamber technique. In this work, it is clear that when $R^2_{adj-non-lin} > R^2_{adj-lin}$, the application of linear regression models is not recommended, as it leads to an underestimation of GHG fluxes by 10 to 50%. This suggests that adopting only or mostly linear regression models will affect the GHG inventories obtained by using closed static chambers. According to our results, the misuse of the usual R^2 parameter and only the linear regression model to estimate the fluxes will lead to reporting erroneous information on the real contribution of GHG emissions from wastewater. Therefore, the R^2_{adj} and non-linear regression model analysis should be used to reduce the biases in flux estimation by the inappropriate application of only linear regression models.

Keywords
Greenhouse Gas Emission, Static Chambers, Stabilization Ponds, Anaerobic Ponds

2.1 INTRODUCTION

Closed static flux chambers have been widely used for measuring greenhouse gas (GHG) emissions from aquatic ecosystems and wastewater treatment systems (Huttunen *et al.*, 2003; Johansson *et al.*, 2004; Lambert and Fréchette, 2005; Singh *et al.*, 2005; Stadmark and Leonardson, 2005; Søvik *et al.*, 2006; Yacob *et al.*, 2006; Søvik and Kløve, 2007; Mander *et al.*, 2008). This technique is widely applied because it has a high degree of adaptability and sensitivity, and is easy to use to simultaneously measure CO_2, CH_4 and N_2O fluxes. Other techniques, such as the eddy co-variance method (Wille *et al.*, 2008), are more complex and expensive and are difficult to use in multiple sites and field conditions (Kroon *et al.*, 2008; Sachs *et al.*, 2008). The closed static chamber technique also has its difficulties which may be associated with design aspects such as height of chamber, chamber area/perimeter ratio, insulation (Rochette and Eriksen-Hamel, 2008), disturbances during measurement, lack of air mixing in the chamber, temperature and under/over pressure within the chamber, as well as air sample handling and storage (Matthews *et al.*, 2003; Vachon *et al.*, 2010).

The closed static chamber technique consists of sealing off a certain volume of air immediately above the water or soil surface (head space) for a period of time of typically 20 to 60 minutes (Smith and Conen, 2004). During this period, the gas concentration in this space increases to a level that can be determined by gas chromatography or infrared analysis. The flux is then calculated from the rate of increase of gas concentration over time within the chamber headspace (Lambert and Fréchette, 2005). This calculation is based on the assumption of a linear increase in the concentration of the different gasses in the head space (Anthony *et al.*, 1995). However, this assumption has been widely applied to GHG emissions from soils with the conclusion that gas exchange may not be constant over time because of the non-steady-state conditions of closed static chambers – and most likely of the natural processes occurring in the soil (Livingston *et al.*, 2006; Kutzbach *et al.*, 2007). The result of this inaccuracy in the basic assumption leads to an underestimation of GHG fluxes. A similar phenomenon may occur when the technique is used for GHG measurements in aquatic systems which may imply for example that studies reporting GHG emissions from wastewater treatment systems may equally be underestimating the GHG.

This study was therefore implemented to assess the validity of linear regression to estimate GHG emissions from stabilization ponds using the static chamber technique by comparing the fluxes obtained from linear and non-linear models. To do this, the analysis of the chamber headspace concentration data from static chambers was based on the comparison of R^2 and $R^2_{adjusted}$ coefficients to determine the goodness- of- fit for linear and non-linear models (i.e. linear, quadratic or exponential).

2.2 MATERIAL AND METHODS

2.2.1 Field conditions

The experiments on GHG measurements were conducted at the anaerobic pond (AP) of a full-scale waste stabilization pond (WSP) system. The WSP is located in the experimental research station for wastewater treatment and reuse in Ginebra, a small town of 10,000 inhabitants located in south-west Colombia (3°43'25.98 N, 76°15'59.45 W), at an altitude of 1040 MASL. The average ambient temperature at the site is 26 °C.

The AP influent is exclusively from domestic sources and reaches the AP after passing through a fine screen to remove coarse material. The design characteristics of the AP are: flow rate 864 $m^3 d^{-1}$, depth 4.0 m, and theoretical hydraulic retention time of 2 days. The effluent from the AP is transferred to a secondary facultative pond.

2.2.2 Closed chamber technique

The GHG measurements (i.e. CO_2, CH_4 and N_2O) in the AP were based on the closed static chamber technique (also called a transient or non-steady-state system). The criteria to standardize the methodology were similar to those for GHG measurements in soils (Anthony *et al.*, 1995; Hutchinson *et al.*, 2000; Kutzbach *et al.*, 2007; Kroon *et al.*, 2008) and aquatic systems (Huttunen *et al.*, 2003; Matthews *et al.*, 2003; Johansson *et al.*, 2004; Lambert and Fréchette, 2005; Mander *et al.*, 2008; Vachon *et al.*, 2010).

Two similar closed static chambers modified with an air circulation pump (**Figure 2.1**) were constructed to measure CH_4, CO_2, and N_2O fluxes at the pond's surface. The chambers were cylindrical (0.3 m x 0.3 m: diameter x height) and were constructed using 4.5 mm-thick transparent acrylic sheets. The chamber dimensions had an area/perimeter ratio of 75. On top of the chambers two holes were drilled to insert two gas-tight butyl rubber stoppers. One stopper was used to set a thermometer while the other was the sampling port. On the sampling port a 0.30 m plastic tube (PVC, i.d. 3 mm) was attached. The free end of each plastic tube was connected to a three-way stopcock that was used to take samples. A peristaltic pump (Cole Parmer, Masterflex Model Nr. 77521-57, and Barrington, Illinois, U.S.A.) with a flow rate of 75 ml min^{-1} (4.5 l h^{-1}) was connected to establish the circulation of air within the chamber's headspace.

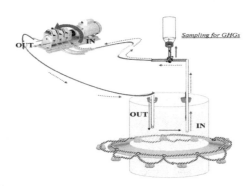

Figure 2.1 Closed static chamber fitted with an air pump to improve air mixing

2.2.3 Sampling protocol

At the beginning of sampling a 70 mm-thick Styrofoam block was added to the rim of the chamber to keep the device floating. Then the chambers were installed gently on the water surface of the AP. The chambers were partially submerged verifying that the edges were about 50 mm beneath the water surface to prevent gas leakage from the chamber. The chambers were anchored with lines to the banks of the pond, to prevent movement and disturbances during the sampling. A boat was used to fix the chambers at the measurement points and the sampling only started once disturbances and turbulence had stopped.

The measurement time was taken as 12 min. due to frequent random gas bubbling events and high productivity of CH_4 and CO_2 as can be expected in AP. This relatively short time provides an adequate balance to detect concentration changes in the chamber and to minimize the disturbances caused by gas ebullition. In addition, based on recommendations by Rochette and Eriksen-Hamel (2008) four discrete air samples were taken during the monitoring time. Thus, CO_2, CH_4 and N_2O samples were taken at 0, 4, 8, and 12 minutes. The fluxes were measured every two weeks at the same period of the day between 12:00 and 14:00 hours. Air samples from the sampling ports of the chambers were taken with 20 ml nylon syringes equipped with three-way valves (**Figure 2.1**). Then the gas sample was taken directly through a needle into a pre-evacuated vial of 5 ml.

2.2.4 Analysis of GHG concentration in samples

CO_2. CO_2 was measured by an infrared spectrophotometer Qubit S151 CO_2 analyser (Loligo Systems, Denmark) using 75 ml min^{-1} air as the mobile phase with the temperature of the injector set equal to the ambient temperature.

CH_4. CH_4 was analysed by gas chromatography (Shimadzu Co., Japan) equipped with a flame ionic detector (FID). The column was Porapak Q (80-100 mesh), 2 m in length and 2 mm in internal diameter, and the temperatures at the injector, column, and detector were 80, 70, and 320 °C, respectively. The flow rate of the carrier gas (N_2) was 22 ml min^{-1}.

N_2O. The N_2O concentration was analysed by means of gas chromatography (Shimadzu Co., Japan) equipped with an electron capture detector (ECD) and a Porapak column Q 80-100 mesh 2m*2mm retention gap, using 22 ml min^{-1} N_2 was the carrier gas, and the injector, column, and detector temperatures were 80, 70, and 320 °C, respectively. The flow rate of the carrier gas (N_2) was 22 ml min^{-1}.

All samples were measured within 2 weeks after sampling. After every 10 samples the system was calibrated. In all cases Scotty standard gases were used for calibration (500 ppmv CO_2, 10 ppmv CH_4, 1 ppmv N_2O).

2.2.5 Evaluation of closed static chambers

A series of studies to evaluate the closed static chambers for GHG measurement from the AP were conducted to (i) determine the comparability of measurements using the closed static chambers technique, (ii) quantify the differences between GHG fluxes from chamber data when linear and non-linear regression are used.

Reproducibility of GHG measurements using the closed static chamber technique
The reproducibility of the closed static chamber technique was evaluated considering: (i) The pattern of change in the GHG concentrations in the chamber headspace over time. The flux estimation is based on concentration data over time, C(t). Since the water-atmosphere GHG exchange is driven primarily by diffusion, then C(t) should generate accumulation curves physically tenable for production and consumption situations. A visual inspection was used in this study to classify the C(t) behaviour related to concentration rises in the chamber headspace. The curves that showed a non-linear trend (i.e. quadratic or exponential) were discarded, whereas the remaining curves were used to calculate the GHG fluxes.

(ii) The reproducibility of GHG measurements obtained using chambers. Two similar transparent chambers were placed directly adjacent to one another on the AP to determine the confidence and reproducibility of GHG measurements using closed static chambers. The Kolmogorov-Smirnov test was used to check for normality of the data. Afterwards, ANOVA and Wilcoxon tests were sequentially applied to data sets so as to compare the differences between the absolute values of concentrations obtained from each chamber.

Flux estimation using linear and non-linear regression
The gas flux depends on the concentration change over time. The CH_4, CO_2 and N_2O fluxes were calculated by Eq. 2.1:

$$F = \frac{dC}{dt_{t=0}} \times \frac{V_c}{A} \times \frac{1440\ min}{d} \quad (Eq.\ 2.1)$$

F= the flux of CH_4, CO_2, N_2O (g m^{-2} d^{-1}); dC/ dt$_{t=0}$ = slope of the gas concentration curve (g m^{-3} min^{-1}); V_c= volume of the chamber (m^3); A=the cross-sectional area of the chamber (m^2).

The flux is directly proportional to the initial slope (dC/dt $_{t=0}$). This initial slope is used as the flux estimate because t_0 is assumed to be the only time when the true exchange rate is unaffected by the presence of the chamber. In Eq. 2.1, the slope (dC/dt $_{t=0}$) was estimated by fitting two different regression models to the concentration measurements by ordinary least-square regression: (1) a linear model and (2) a non-linear model (i.e. quadratic or exponential).

For linear regression, the chamber data were fitted to a straight line given by:

$$C = a_o + a_1 t \quad (Eq.\ 2.2)$$

In which C is the measured concentration at time t, a_0 is the y-axis interception and a_1 the line slope.

For non-linear regression, two models were used to fit the chamber data: *quadratic* and *exponential.* In the case of a quadratic model, the mathematical expression is given by:

$$C = b_o + b_1t + b_2t^2 \quad \text{(Eq. 2.3)}$$

Where b_0, b_1, *and* b_2 are regression parameters derived using the polynomial regression function in Excel 2007. For this model, $(dC/dt_{\ t=0})$ is equal to parameter b_1. Parameter b_0 is the air concentration at time=0 and b_2t^2 can be regarded as an extra loss term as compared to the linear regression (Wagner *et al.* 1997). In addition, the slopes of this model attained the calculation of fluxes under the constraint that $(dC/dt_{\ t=0})\ /\ (dC/dt_{\ t=12}) > 0$ based on recommendations by Stolk *et al.* (2009).

For the exponential model the concentration behaviour was fitted to the expression based on Fick´s first law given by:

$$C = C_{max} - (C_{max} - C_o).e^{-kt} \quad \text{(Eq. 2.4)}$$

In which C_{max} is the maximum concentration that can be reached in the chamber, C_0 is the concentration at t=0, and k is a rate constant. In this case, the slope is given by:

$$\frac{dC}{dt} = k.(C_{max} - C_o) \quad \text{(Eq. 2.5)}$$

The parameters for each model, their goodness of fit, the coefficient of determination (R^2), and the adjusted coefficient of determination R^2_{adj} were estimated using the library *curve fitting application* of MATLAB (MathWorks Inc, 2011, Version 7.12.0.635). A preliminary selection of best-fitted models was done under the $R^2 > 0.85$ criterion which was reported previously by other authors (Huttunen *et al.*, 2002; Lambert and Fréchette, 2005). However, the definitive model to estimate the fluxes was selected based on the highest R^2_{adj} value of each model (i.e. linear, quadratic or exponential).

2.3 RESULTS

2.3.1 Reproducibility of greenhouse gas measurements using closed static chambers

Figure 2.2 shows the different patterns that were obtained for the CH_4, CO_2 and N_2O chamber data in this study. Only patterns similar to those in Figs. 2.2a, 2.2c, and 2.2e were taken into account to determine the CH_4, CO_2 and N_2O fluxes. Patterns shown in Figs. 2.2b, 2.2d, and 2.2f could not be given a plausible explanation and were discarded. The percentages of curves accepted were 93.6, 87.2, and 73% for CH_4, CO_2 and N_2O, respectively ($n = 47$). This indicates that for CH_4, and CO_2 in this study the number of curves with non-plausible explanations was relatively low, while N_2O showed the highest number of rejected curves.

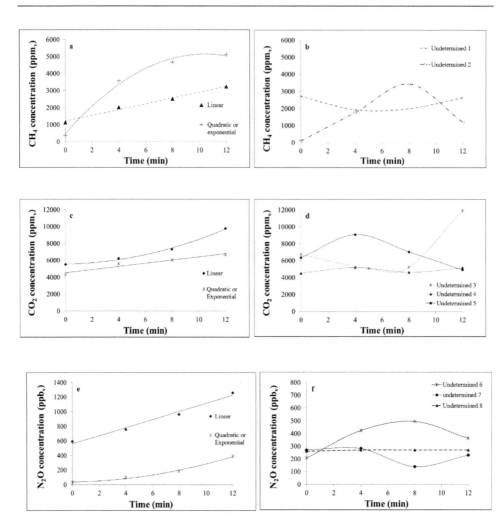

Figure 2.2 Typical curves showing the different behaviours of GHG concentration over time observed in the headspace of chambers during the measurement time. (a) Curves with plausible explanation for CH_4 flux that fits to linear, quadratic or exponential regression, (b) Curves with no plausible explanation for CH_4 due to disturbances in the chamber during measurement, (c) curves plausibly explainable for CO_2 flux that fits to linear or non-linear regression, (d) curves with no plausible explanation for CO_2 flux, (e) curves explaining N_2O flux that fits to either linear or non-linear regression, and (f) curves with no plausible explanation for N_2O flux.

Figure 2.3 shows the CH_4 and CO_2 concentrations measured during the comparability-evaluation sampling campaign. According to Kruskal-Wallis statistical tests, there were no significant significant differences between concentrations measured in both chambers (p =

0.3749). This suggests that the closed static chamber technique under prescribed similar conditions (i.e. time
of day, similar chamber, measurement time and pond) yields good precision and comparable results for GHG measurements.

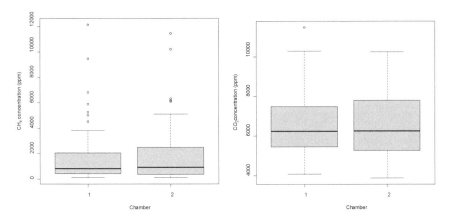

Figure 2.3 Distribution of CH_4 and CO_2 concentrations measured by equal chambers placed in adjacent positions.

In **Figure 2.4**, the CO_2 and CH_4 concentrations are plotted against the measurement time for each chamber, respectively. The blue line represents the mean profile obtained at each chamber for both gases. According to this, the headspace concentrations were different ($p<0.05$) at each time (0, 4, 8, and 12 min) for both CO_2 and CH_4. This suggests a rate of increase of gas concentration with time within the chamber headspace, and the maximum values were reached at t =12 min. A similar behaviour was observed for the CH_4 and CO_2 mean profiles between chambers, which confirmed the reproducibility of greenhouse gas measurements obtained from the static chamber technique. Furthermore, the observed CO_2 and CH_4 concentrations measured at t = 0 showed less dispersion than data for the other sampling times (n = 13). The latter indicates that conditions change within the chambers as time passes, and this may be attributed to the dynamics of AP or other external influences on the chamber headspace.

2.3.2 Flux calculation using linear and non-linear regression models
Model selection to calculate the fluxes
Table 2.1 shows the goodness of fit for three regression methods based on the R^2 and R^2_{adj} criteria. In general, the R^2 quadratic coefficients of CO_2, CH_4 and N_2O data concentrations were higher than those for linear or exponential models. This shows that under the R^2 criterion, the quadratic model has the best fit. Additionally, the R^2 values from linear and exponential regressions suggest that in general the linear model has a better fit than the exponential one.

However, when the R^2_{adj} criterion was used the percentage of data best fitted to the linear model was increased and reached a similar percentage to that of quadratic regression.

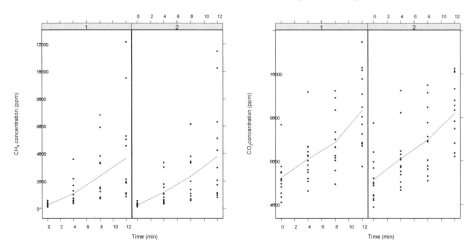

Figure 2.4 Mean profiles for CH_4 and CO_2 concentrations measured by equal chambers placed in adjacent positions to determine the reproducibility of the closed static chamber technique. A 12-min measurement period was used, and the samples were taken at 0, 4, 8, 12 min.

Table 2.1 Goodness-of- fit for the linear, quadratic and exponential regression models as compared by the R^2 and R^2_{adj} criteria.

Gas		$R^2_{Lin}>R^2_{quad}$	$R^2_{Lin} > R^2_{exp}$	$R^2_{adjLin}> R^2_{adjquad}$	$R^2_{adjLin}> R^2_{adjexp}$
	n_{tot}	% n_{total}	% n_{total}	% n_{total}	% n_{total}
CO_2	41	2.4	56.0	43.9	75.6
CH_4	44	0.0	56.8	54.6	84.8
N_2O	34	0.0	35.3	55.9	61.9

Comparison of initial slopes and fluxes
In **Figure 2.5**, CH_4, CO_2 and N_2O initial linear slopes ($dC/dt_{t=0}$) are plotted against their quadratic and exponential counterparts. In general, for the three gases studied when $R^2_{adj-lin} > R^2_{adj-cuad}$, the linear and quadratic initial slopes were similar or relatively close to 1:1 (Fig.2.5a, b, e). On the other hand, when the CO_2, CH_4, and N_2O data were best fitted to a quadratic regression model ($R^2_{adj-cuad} > R^2_{adj-lin}$), the initial quadratic slopes differed considerably when compared to the linear ones. For instance, in the case of CH_4, in 70% of all measurements which were best fitted to a quadratic regression the initial slopes were lower than those from the linear model (i.e. quadratic initial slopes close to 80 ppmv.min^{-1} could be half or less than their linear counterparts). In contrast, for the remaining 30% of the data the quadratic initial slopes could be up to twice the initial slope of linear regression.

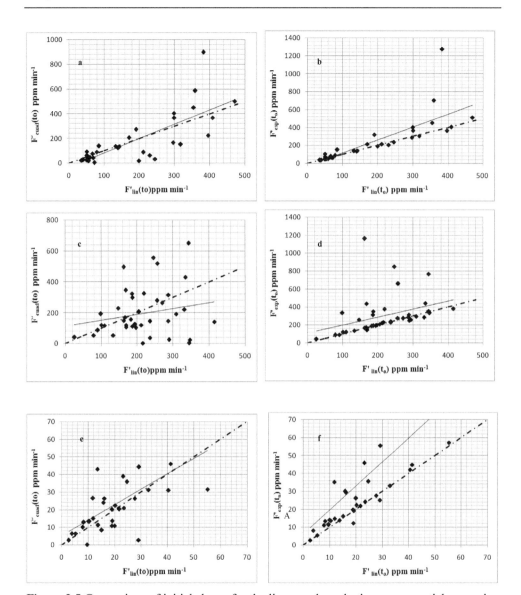

Figure 2.5 Comparison of initial slopes for the linear and quadratic or exponential regression curves. (a) CH_4 linear versus quadratic, (b) CH_4 linear versus exponential, (c) CO_2 linear versus quadratic, (d) CO_2 linear versus exponential, (e) N_2O linear versus quadratic, (f) N_2O linear versus exponential. The dashed line represents the 1:1 ratio.

An analysis of the line curvature based on the ratio $(dC/dt_{\ t=0})/(dC/dt_{\ t=12})$ was done so as to determine which initial slope is to be used to estimate the flux. Thus, it was found that the initial quadratic slopes were lower than the linear ones when this ratio was below 0.2. This in turn is related to a large concavity of the curve. Nonetheless, when this ratio was close to 1.0, both quadratic and linear initial slopes were similar, and the concavity of the curve was small. On the other hand, for a ratio close to 5.0, the initial quadratic slopes were twice as high as the linear ones and the curves were convex. Since the expression represented by Eq. 2.1 follows the diffusion model, only the convex curves or those with small concavities are plausible from a physical standpoint. Accordingly, the initial quadratic slopes, which were significantly different from the linear ones, should not be taken into account for the estimation of fluxes. Likewise, the ratios $(dC/dt_{t=0}) / (dC/dt_{t=12})$ larger than 5.0 for the initial quadratic slopes should also be discarded.

When the linear model was the best-fitting model compared to the exponential ($R^2_{adjlin} > R^2_{adjexp}$) model, the initial slopes of the two models were relatively close to a 1:1 ratio. On the other hand, the initial slopes for the curves that were best fitted to an exponential ($R^2_{exp} > R^2_{adjlin}$) model were larger than the initial linear slopes. For CH_4 data ($n_{exp} = 4$) that were fitted to the exponential model, the initial slopes were 0.8 to 3.0 times larger than the initial linear slopes (e.g. Fig. 2.5b). Similar results were obtained for CO_2 and N_2O where the initial exponential slopes were twice those for linear ones (Fig. 2.5d, Fig. 2.5f). However, the largest initial exponential slopes were discarded for estimating the fluxes because the parameters of the regression based on Eq. 2.4 were unrealistic and physically unattainable (i.e. when the C_o parameter of exponential regression showed a significant difference compared to the concentration measured at time zero (t=0).

According to the criteria mentioned above, it was necessary to recalculate the distribution of data that were best fitted to a linear regression, quadratic or exponential model (**Table 2.2**). In general, for CO_2, CH_4, and N_2O the majority of measurements were well represented by the linear regression model. However, approximately 40% of the data were best fitted to a non-linear regression model (quadratic or exponential). This result reconfirms that not only a linear regression model approach should be used to estimate the GHG fluxes when using the closed static chamber technique.

The initial slopes of the regression functions are directly proportional to the GHG flux at the beginning of chamber operation $F_{net}(t_0)$, which is considered to be the best estimate of the undisturbed flux just before chamber closure (Forbrich et al. 2010). Since the initial quadratic or exponential slopes were higher than the linear ones (when $R^2_{adj-non-lin} > R^2_{adj-Lin}$), the estimated GHG fluxes when assuming linear behaviour causes an underestimation of the fluxes.

This can be tested by calculating the fluxes to concentration data that showed a coefficient $R^2_{adj-exp} > R^2_{adj-Lin}$. For instance, taking the CO_2 data shown in Fig. 2.5d, the initial slope considering linear regression was 343.9 ppmv min^{-1}, while the initial slope under the exponential model was 766.9 ppmv min^{-1}. Thus, in this example, the CO_2 fluxes estimated by linear and exponential regression models were 232.5 and 519 g m^2 d^{-1}, respectively. In consequence, the

CO_2 flux estimated by linear regression is only 44.7% of the flux estimate by exponential regression. In general, when $R^2_{adj-non-lin} > R^2_{adj-Lin}$ if only the linear regression is considered for CO_2, CH_4, and N_2O, then the calculated fluxes showed an underestimation between 10 to 50 % of the flux estimated by the non-linear regression model (i.e. quadratic or exponential).

Table 2.2 Best-fitting data to linear, quadratic, or exponential regression models to estimate GHG fluxes

Gas		Linear	Quadratic	Exponential
	n_{Total}	% n_{Total}	% n_{Total}	% n_{Total}
CO_2	35	65	20	15
CH_4	31	60	16	24
N_2O	30	60	17	23

2.4 DISCUSSION

2.4.1 Reproducibility of greenhouse gas measurements using the closed static chamber measurement technique

The sampling campaigns provided an assessment of comparability of the closed static chamber technique to estimate GHG fluxes (i.e. CH_4, CO_2, and N_2O in the AP). This is consistent with the use of this technique in other studies (Huttunen *et al.*, 2003; Johansson *et al.*, 2004; Lambert and Fréchette, 2005; Singh *et al.*, 2005; Stadmark and Leonardson, 2005; Liikanen *et al.*, 2006; Yacob *et al.*, 2006; Søvik and Kløve, 2007; Mander *et al.*, 2008). However, in these studies a data analysis for gas concentration behaviour within the chamber headspace was not shown, nor was the comparability of the respective measurements reported.

An intrinsic limitation of using closed static chambers is that not all data of gas concentrations measured within a chamber headspace can be used to estimate the flux. However, the percentage of curves rejected in this study for CH_4 and CO_2 flux estimation was lower than reported by Lambert and Fréchette (2005) in a hydroelectric reservoir and similar to results reported by Matthews *et al.* (2003) in sheltered aquatic surfaces. The number of curves rejected for N_2O was lower than reported by Johansson *et al.* (2003).

Rejections had two main reasons; either the N_2O initial concentration was lower than the ambient concentrations (background) at $t= 0$ or there was no variation between the sample concentrations in time. A measurement at $t= 0$ of a concentration significantly different to the background could be attributed to leakage or disturbances during the deployment of the chamber. The unobserved variation of concentration with time probably indicates that during the measurement time, the rate of N_2O production was low and constant, which leads to non-detectable or very small fluxes. This was also observed in studies carried out on soils where rejection of around 55% (Kroon *et al.*, 2008) and 75% (Stolk *et al.*, 2009) of N_2O curves has been reported .

The duplicate measurements performed in two adjacent chambers demonstrated the reproducibility of data obtained from closed static chambers. The low variability of data obtained between two simultaneous measurements suggests that chambers were properly designed, the handling and storage of samples were appropriate, and disturbances during the deployment were minimized. The 75 mm ratio Area/Perimeter, the 4 samples undertaken during the measurement time, the use of glass vials for sample store and handling, and the deployment time were according to the requirements suggested by Rochette and Eriksen-Hamel (2008) to consider the reliability of the closed static chamber technique.

2.4.2 Model selection and flux estimation

In most studies of GHG measurements in aquatic or wastewater treatment systems, the evaluation of gas concentration behaviour within a chamber headspace has been based on linear regression and the R^2 criterion. However, according to the results obtained in this study the goodness of fit of data should instead be evaluated using the R^2_{adj} criterion. Additionally, when the data are best-fitted to a non-linear regression model, a curvature analysis of the correlation line must be done.

According to the R^2 coefficients obtained, the data were best fitted to a quadratic regression model instead of the customary linear model. In many polynomial regression models, the addition of parameters to the correlation equation increases the R^2 value. Thus, when a model includes several parameters, it can easily fit further noise or disturbed measurements. This means that considering only the R^2 coefficient as the best-fitting criterion to select the model may actually be incorrect. By contrast, the R^2_{adj} coefficient may be more appropriate to determine the goodness of fit, as shown in this study by the higher % of data that were best fitted to the linear compared to the quadratic regression model when R^2_{adj} was the selection criterion (**Table 2.1**). R^2 always varies between 0 and 1 for the polynomial regression models that the basic fitting tool generates. However, R^2_{adj} for some models can be negative, indicating that such models include too many terms. Therefore, R^2_{adj} is a more appropriate criterion compared to the usual R^2.

When experimental data are best fitted to quadratic or exponential regression models, an analysis of parameters should be done in addition to the R^2_{adj} criterion. The parameters should remain within an acceptable range to describe the model represented by Eq. 2.1. For instance, some data that are better fitted to quadratic regression models than linear ones exhibited a pronounced concavity. Such concavity is related to the ratio $(dC/dt_{t=0}) / (dC/dt_{t=12})$. If this ratio is negative or close to zero, the quadratic curve regression has a negative concavity and the slope tends to decrease. On the contrary, for larger ratios the quadratic curve regression has a positive concavity and the slope may increase. Owing to the fact that an excessive curvature (either positive or negative) of the quadratic regression model does not contribute to explain the physical diffusion phenomenon represented by Eq. 2.1, the initial slopes could provide an erroneous under or over-estimation of the flux. Consequently, the quadratic model becomes

limiting and probably linear and exponential models should be considered instead to get a better CH_4, CO_2 and N_2O flux estimation. In the case of the exponential model there is a high probability that experimental data will fit to this model due to the number of parameters, but however, the proper physical explanation in some cases could be questionable. For instance, the parameters C_o, and C_{max} predicted by the regression equation should be similar to the ones measured analytically in the field or in the lab. Thus, large differences between these values suggest that selecting the exponential regression model in this case is not the best method to estimate the flux.

In most studies of GHG emissions from aquatic and wastewater treatment systems, the linear regression model is selected to estimate the fluxes. Although the linear model provides simplicity, in some cases the concentration data of the chamber deviate from linear behaviour. In this study, the results indicated that the linear model might be applied to estimate fluxes only in some instances (60% data). However, a significant non-linearity was found in some data sets and this suggests that different models other than the linear regression should be considered to estimate GHG fluxes. The non-linearity trend and consequent underestimation of fluxes when using only linear regression have been analysed in several prior studies related to GHG emissions from soils (Wagner *et al.*, 1997; Kutzbach *et al.*, 2007; Kroon *et al.*, 2008; Stolk *et al.*, 2009; Forbrich *et al.*, 2010). However, for GHG emissions from wastewater treatment systems this has not yet been reported. From this work, it is clear that when $R^2_{adj-non-lin} > R^2_{adj-lin}$, the application of linear regression models is not appropriate and may lead to an underestimation of GHG fluxes of between 10 to 50%. Therefore, the incorrect use of the usual R^2 parameter and only the linear regression model to estimate the fluxes may lead to severe underestimation of the real contribution of GHG emissions from wastewater.

2.5 CONCLUSIONS

The flux data suggest that closed static chamber is a good analytical methodology to estimate GHG emissions from wastewater stabilization ponds (i.e. AP). However, GHG flux estimation using closed static chambers poses potential error sources related to water surface disturbances, temperature, concentration, and pressure gradients within the chamber, moisture saturation and lack of mixing in the headspace. These error sources should be minimized to prevent a high number of gradient concentration curves being rejected.

The linear regression model has been widely used to estimate GHG fluxes from aquatic and wastewater treatment systems. In addition, the coefficient R^2 is the only criterion used to decide the data concentration goodness of fit when measured within the headspace of the chamber.

However, the results of this study showed that in some cases the experimental data displayed a significant non-linear trend, thus affecting the flux calculation. Every time the data sets were fitted to the best non-linear model, but the flux was still calculated by the initial linear slope, then an underestimation of the actual flux was obtained. In general, the underestimation ranged between 10 and 50 %. We thus suggest the use of the R^2_{adj} criterion because it detects the influence of extra parameters in the regression models.

2.6 ACKNOWLEDGEMENTS

This study has been carried out within the framework of the European research project SWITCH (Sustainable Urban Water Management Improves Tomorrow's City's Health). SWITCH is supported by the European Commission under the 6th Framework Programme and contributes to the thematic priority area of 'Global Change and Ecosystems' [1.1.6.3] Contract Nr. 018530-2. The author is also indebted to ACUAVALLE ESP and the Universidad del Valle for their support. Our recognition also goes to undergraduate students in sanitary engineering John Camacho and José Reyes, for their assistance.

2.7 REFERENCES

Anthony, W.H., Hutchinson, G.L. and Livingston, G.P. (1995). Chamber measurement of soil-atmosphere gas exchange: linear vs. diffusion-based flux models. Soil Science Society of America Journal 59(5), 1308-1310.

Forbrich, I., Kutzbach, L., Hormann, A. and Wilmking, M. (2010). A comparison of linear and exponential regression for estimating diffusive CH_4 fluxes by closed-chambers in peatlands. Soil Biology and Biochemistry 42(3), 507-515.

Hutchinson, G.L., Livingston, G.P., Healy, R.W. and Striegl, R.G. (2000). Chamber measurement of surface-atmosphere trace gas exchange: numerical evaluation of dependence on soil, interfacial layer, and source/sink properties. Journal of Geophysical Research: Atmospheres 105(D7), 8865-8875.

Huttunen, J.T., Alm, J., Liikanen, A., Juutinen, S., Larmola, T., Hammar, T., Silvola, J. and Martikainen, P.J. (2003). Fluxes of methane, carbon dioxide and nitrous oxide in boreal lakes and potential anthropogenic effects on the aquatic greenhouse gas emissions. Chemosphere 52(3), 609-621.

Huttunen, J.T., Väisänen, T.S., Heikkinen, M., Hellsten, S., Nykänen, H., Nenonen, O. and Martikainen, P.J. (2002). Exchange of CO_2, CH_4 and N_2O between the atmosphere and two northern boreal ponds with catchments dominated by peatlands or forests. Plant and Soil 242(1), 137-146.

Johansson, A.E., Gustavsson, A.M., Öquist, M.G. and Svensson, B.H. (2004). Methane emissions from a constructed wetland treating wastewater--seasonal and spatial distribution and dependence on edaphic factors. Water Research 38(18), 3960-3970.

Johansson, A.E., Klemedtsson, Å.K., Klemedtsson, L. and Svensson, B.H. (2003). Nitrous oxide exchanges with the atmosphere of a constructed wetland treating wastewater. Parameters and implications for emissions factors. Tellus 55B(3), 737-750.

Kroon, P.S., Hensen, A., van den Bulk, W.C.M., Jongejan, P.A.C. and Vermeulen, A.T. (2008). The importance of reducing the systematic error due to non-linearity in N_2O flux measurements by static chambers. Nutrient Cycling in Agroecosystems 82(2), 175-186.

Kutzbach, L., Schneider, J., Sachs, T., Giebels, M., Nykänen, H., Shurpali, N.J., Martikainen, P.J., Alm, J. and Wilmking, M. (2007). CO_2 flux determination by closed-chamber methods can be seriously biased by inappropriate application of linear regression. Biogeosciences Discussions 4(4), 2279-2328.

Lambert, M. and Fréchette, J.L. (2005). Greenhouse Gas Emissions - Fluxes and Processes. Tremblay, A., Varfalvy, L., Roehm, C. and Garneau, M. (eds), pp. 37-60, Springer, Berlin.

Liikanen, A., Huttunen, J.T., Karjalainen, S.M., Heikkinen, K., Vaisanen, T.S., Nykanen, H. and Martikainen, P.J. (2006). Temporal and seasonal changes in greenhouse gas emissions from a constructed wetland purifying peat mining runoff waters. Ecological Engineering 26(3), 241-251.

Livingston, G.P., Hutchinson, G.L. and Spartalian, K. (2006). Trace Gas Emission in Chambers: A Non-Steady-State Diffusion Model. Soil Science Society of America Journal 70(5), 1459-1469.

Mander, Ü., Lõhmus, K., Teiter, S., Mauring, T., Nurk, K. and Augustin, J. (2008). Gaseous fluxes in the nitrogen and carbon budgets of subsurface flow constructed wetlands. Science of the Total Environment 404(2-3), 343-353.

Matthews, C.J.D., Louis, V.L.S. and Hesslein, R.H. (2003). Comparison of three techniques used to measure diffusive gas exchange from sheltered aquatic surfaces. Environmental science & technology 37(4), 772-780.

Rochette, P. and Eriksen-Hamel, N.S. (2008). Chamber measurements of soil nitrous oxide flux: are absolute values reliable? Soil Science Society of America Journal 72(2), 331-342.

Sachs, T., Wille, C., Boike, J. and Kutzbach, L. (2008). Environmental controls on ecosystem-scale CH4 emission from polygonal tundra in the Lena River Delta, Siberia. Journal of geophysical research 113(G3), G00A03.

Singh, V.P., Dass, P., Kaur, K., Billore, S.K., Gupta, P.K. and Parashar, D.C. (2005). Nitrous oxide fluxes in a tropical shallow urban pond under influencing factors. Current Science 88(3), 478.

Smith, K.A. and Conen, F. (2004). Impacts of land management on fluxes of trace greenhouse gases. Soil Use and Management 20(2), 255-263.

Søvik, A.K., Augustin, J., Heikkinen, K., Huttunen, J.T., Necki, J.M., Karjalainen, S.M., Kløve, B., Liikanen, A., Mander, Ü., Puustinen, M., Teiter, S. and Wachniew, P. (2006). Emission of the greenhouse gases nitrous oxide and methane from constructed wetlands in Europe. Journal of Environmental Quality 35(6), 2360-2373.

Søvik, A.K. and Kløve, B. (2007). Emission of N_2O and CH_4 from a constructed wetland in southeastern Norway. Science of the Total Environment 380(1), 28-37.

Stadmark, J. and Leonardson, L. (2005). Emissions of greenhouse gases from ponds constructed for nitrogen removal. Ecological Engineering 25(5), 542-551.

Stolk, P.C., Jacobs, C.M.J., Moors, E.J., Hensen, A., Velthof, G.L. and Kabat, P. (2009). Significant non-linearity in nitrous oxide chamber data and its effect on calculated annual emissions. Biogeosciences Discuss 6, 115-141.

Vachon, D., Prairie, Y.T. and Cole, J.J. (2010). The relationship between near-surface turbulence and gas transfer velocity in freshwater systems and its implications for floating chamber measurements of gas exchange. Limnology and Oceanography 55(4), 1723-1732.

Wagner, S.W., Reicosky, D.C. and Alessi, R.S. (1997). Regression models for calculating gas fluxes measured with a closed chamber. Agronomy journal 89(2), 279-283.

Wille, C., Kutzbach, L., Sachs, T., Wagner, D. and Pfeiffer, E.V.A. (2008). Methane emission from Siberian arctic polygonal tundra: eddy covariance measurements and modeling. Global Change Biology 14(6), 1395-1408.

Yacob, S., Ali Hassan, M., Shirai, Y., Wakisaka, M. and Subash, S. (2006). Baseline study of methane emission from anaerobic ponds of palm oil mill effluent treatment. The Science of the Total Environment 366(1), 187-196.

Chapter 3

Greenhouse Gas Emissions from an Anaerobic Pond Treating Domestic Wastewater under Tropical Climatic Conditions

Abstract

Greenhouse gas emissions from a full-scale anaerobic pond (AP) used for the treatment of domestic sewage were studied under tropical conditions. The GHG fluxes were measured using the static chamber technique. The results showed that the AP was a source of CH_4, CO_2 and N_2O emissions. CH_4 emissions ranged from 13.4 to 178.7 $l.m^{-2}.d^{-1}$, CO_2 from 9.3 to 130.5 $l.m^{-2}.d^{-1}$, while N_2O emissions ranged between 0.0016 and 0.013 $l.m^{-2}.d^{-1}$. According to the average fluxes, the emission rates into the atmosphere for CH_4 and CO_2 were 0.24 m^3 $CH_4.kg$ COD_{rem}^{-1} and 0.18 m^3 $CO_2.kg$ COD_{rem}^{-1}, respectively. A COD mass balance calculation indicated that 37% of the influent COD was converted to CH_4 and 36% left the AP with the effluent. The rest of the COD was accounted for as volatile solids (3.5 %), CH_4 dissolved in the effluent (2.5%) and VSS in sludge settlement (21%). Finally, the GWP total for the AP studied was 37.8 $kgCO_{2eq}.PE^{-1}.y^{-1}$ or 2.18 $kgCO_{2eq}.kgCOD_{rem}^{-1}$, which suggest that operating anaerobic ponds without considering the capture and reuse of biogas, may contribute to global warming.

Keywords
Anaerobic Pond, Greenhouse gases, Biogas, Stabilization Ponds, Wastewater Treatment

3.1 INTRODUCTION

The growing awareness of the impact of greenhouse gases (GHG) from human activities on climate change, triggered the need to identify and quantify the main sources of these gases. Human activities such as agriculture, the production and use of fossil fuels, industrial and agricultural activities, and waste management have increased greenhouse gas concentrations in the past 200 years (El-Fadel and Massoud, 2001). According to Tarasova et al. (2016), the pre-industrial levels, prior to 1750, of CO_2, CH_4 and N_2O had risen by 2011 to 397.7 ppmv, 1833 ppbv, and 327 ppbv, respectively, this means an increase of approximately 43%, 154%, and 21%, respectively. Methane and nitrous oxide are potent greenhouse gases with a global warming potential 28 and 296 times greater than CO_2, respectively (Myhre et al., 2013).

Different researches have determined that wastewater treatment systems are potential sources of anthropogenic GHG emissions contributing to climate change and air pollution (Shahabadi et al., 2009; Sweetapple et al., 2014). Wastewater treatment plants contribute to greenhouse gas emission through the production of CH_4, CO_2 and N_2O from treatment processes, with CO_2 also produced from the energy required for treatment (Cakir and Stenstrom, 2005; Foley et al., 2015). From 1990 to 2025, worldwide CH_4 and N_2O emissions from wastewater are estimated to increased, between 352 and 477 $MtCO_{2eq}$ and between 82 to 100 $MtCO_{2eq}$, respectively (EPA, 2012; Karakurt et al., 2012). This increase in GHG emissions will come mainly from developing countries of East and South Asia, the Middle East, the Caribbean, and Central and South America, mainly due to population increase (Bogner et al., 2007).

However, the figures above mentioned exhibit large uncertainty resulting from the lack of reliable data, which are often inconsistent or incomplete. This fact limits the possible mitigation and reduction strategies that must be implemented to prevent the impact of the wastewater treatment sector on the global warming. Thus, the only effective option to reduce this gap is by developing programs that involve field measurement to estimate the generation of GHG emissions from WWT management (Bogner *et al.*, 2007).

Anaerobic ponds (AP) have been identified as an effective first stage treatment process in wastewater stabilization ponds (WSPs) systems (Mara and Pearson, 1998). The main advantage of a WSPs system with a deep AP, is that it takes up much less area resulting in a considerable reduction in project expenditure (DeGarie *et al.*, 2000; Picot *et al.*, 2003). In an AP, the organic matter is transformed biochemically into biogas (CH_4 and CO_2 mainly). This biogas, if collected, can be used to produce electricity and sequester carbon (Lettinga, 1995; Shilton *et al.*, 2008; Konaté *et al.*, 2013). However, most APs in developing countries do not have a biogas collection system, resulting in the direct emission of greenhouse gases such as CH_4 and CO_2.

The production and emission of CH_4 and CO_2 as greenhouse gases from APs under Sahelian, temperate and subtropical conditions have been reported (Toprak, 1995; Picot *et al.*, 2003; DeSutter and Ham, 2005; Yacob *et al.*, 2005; Konaté *et al.*, 2013). These authors stated that local climatic conditions and COD removal rates are identified as the most important factors affecting GHG from APs. Because the digestion rates of anaerobic microorganisms are largely influenced by temperature (Lettinga *et al.*, 2001), a decrease in both water and air temperature will affect the biogas production (Picot *et al.*, 2003). On the other hand, fluctuations of the COD loading rate may cause irregularities in the daily volume of biogas produced (Toprak, 1995; Konaté *et al.*, 2013). Finally, none of these reported studies provided N_2O data from the AP.

Because APs are an appropriate technology, widely used for wastewater treatment globally, there is a necessity for basic research that provides answers on the impacts related to the GHG emission from this technology. Besides, previous studies did not include N_2O measurements from APs. Therefore, considering that GHG estimation from anaerobic ponds under tropical conditions appears to have been largely overlooked, the aim of this study was (i) to measure CH_4, CO_2 and N_2O greenhouse gas emissions from a full scale anaerobic pond treating domestic wastewater, and, (ii) to investigate the influence of temperature, organic load and sludge accumulation on CH_4, CO_2 and N_2O emissions.

3.2 METHODOLOGY

3.2.1 Site description
CH_4, CO_2, and N_2O fluxes were determined for a full-scale anaerobic pond (AP) during a 7-month period. The AP is part of a WSPs treatment system located in Ginebra (Colombia), a small town of 10,000 inhabitants (3°43'25.98 N, 76°15'59.45 W). The climate of Ginebra, at

an altitude of 1040 m above sea level, is tropical with a mean annual ambient air temperature of 26°C.

The WSPs was constructed in 1993 and has been used for the treatment of domestic wastewater. The AP was designed with a length and width of 52 and 26 m, respectively. The AP treats a flow rate of 1728 $m^3.d^{-1}$ and a maximum holding volume of 3438 m^3, with a maximum pond depth of 4 m, and a theoretical hydraulic retention time of 2 days. Wastewater collected by the main interceptor is passed through a fine screen and grit chamber to remove coarse and sandy materials. Then the raw wastewater is pumped into the AP near the bottom and mixes with the suspended active microbial solids in the sludge blanket. Effluent leaves the pond via a weir, located on the opposite end of the pond and enters into a facultative pond by a pipe.

3.2.2 Wastewater characterization and sludge accumulation

The characteristics of the influent and effluent were determined weekly. During the sampling campaigns, composite 24 h samples for the influent and individual samples for the effluent were collected and analysed. Biochemical oxygen demand (BOD_5), chemical oxygen demand (COD), and total suspended solids (TSS) were measured according to Standard Methods (APHA, 2005). Temperature, pH, and oxidation-reduction potential were measured using electrodes.

The procedure for the determination of effective pond depth was conducted according to the towel test (Mara and Pearson, 1998). Sludge depth in the AP was determined for 45 points evenly distributed over the pond surface, using 5mx5m grids covering the entire pond. The sludge depth sampler consisted of a 6 m piece of 32mm-diameter aluminium rod. Three-dimensional surfaces profiles of the sludge distribution in AP were created with a surface tool available in Mat lab software. This software was also used to calculate the total sludge and water volume for AP. The apparent sludge accumulation rates (m^3/person/yr) were calculated by dividing the total sludge volume by the population of Ginebra and the number of years of operation.

3.2.3 Greenhouse gas measurement

Greenhouse gas emissions (CH_4 and CO_2) from the AP were measured weekly for seven months. The samples were taken from three different locations on the pond water surface: 7, 28 (centre), and 49 m from the pond inlet of the AP, in order to obtain an estimate of GHG emissions.

The technique used to determine GHG emissions was based on the static chambers technique described in the chapter 2 and by Silva *et al.* (2015). Using a special syringe, CH_4, CO_2 and N_2O sampling were taken over a 45-minute period at 15-minute intervals (0, 15, 30, 45 min) from a sampling port a top each chamber. The relatively insignificant 15-mL sample volume did not affect the concentration build up in the static chamber.

CO₂. CO₂ was measured by an infrared spectrophotometer (Qubit S151 Loligo Systems, Denmark) using 75 ml min⁻¹ air as the mobile phase with the temperature of the injector set equal to the ambient temperature.

CH₄. CH₄ was analysed by gas chromatography (Shimadzu Co., Japan) equipped with a flame ionic detector (FID). The Porapak Q column (80-100 mesh), was 2 m long and 2 mm in internal diameter. The temperatures at the injector, column, and detector were 80, 70, and $320°$ C, respectively. Carrier gas (N_2) flow rate was 22 ml min⁻¹.

N₂O. N₂O concentration was similarly analysed by gas chromatography except that the equipment was fitted with an electron capture detector (ECD).

All samples were measured within 2 weeks after sampling. After every 10 samples the system was calibrated. In all cases Scotty standard gases were used for calibration (500 ppmv CO_2, 10 ppmv CH_4, 1 ppmv N_2O).

The CH_4, CO_2 and NO_2 concentrations obtained were processed to calculate net fluxes from linear and non-linear changes in the gas concentration in the chamber headspace according to the method suggested by (Silva _et al._ 2015):

$$ F = \frac{dC}{dt_{t=0}} \times \frac{V_c}{A} \times \frac{1440 \text{ min}}{d} \quad \text{(Eq. 3.1)} $$

_F= the flux of CH₄, CO₂, N₂O (gm⁻²d⁻¹); dC/ dt₁₋₀ = slope of the gas concentration curve (gm⁻³min⁻¹); V_c= volume of the chamber (m³); A=the cross-sectional area of the chamber (m²)._

Statistical analyses were done with SPSS® software (v. 15.0 for Windows). The nonparametric statistical test (_i.e._ Mann-Whitney _U_-test), was used to determine whether the behaviour of the AP in terms of final water quality and GHGs emissions, was significantly different at a significance level α=0.05.

3.3 RESULTS

3. 3.1 Wastewater characteristics and removal efficiencies
The influent COD and TSS values (**Table 3.1**) indicated that wastewater from the municipality of Ginebra can be classified as domestic wastewater with medium strength (Metcalf and Eddy, 2003). The AP was operated with an average volumetric load of 0.120 kgBOD₅.m⁻³.d⁻¹. The maximum water temperature of 25.8°C observed was typical of tropical conditions.

According to the measurements of wastewater characteristics in the AP, the average removal efficiency of COD was 64% which was lower than found in the literature by Peña _et al._ (2000), and at the lower range of the 60-80% removal expected in AP operating under tropical conditions (Mara, 2004). Further, the average of TSS removal was 69%, which indicates that the AP functioned particularly efficiently in the settling of the solids. Finally, the pH values measured in the AP varied between 6.5 and 7.2 being quite stable in the water column (**Table 3.1**).

Table 3.1 Wastewater characteristics (mean ± SD; n=32) for the influent and effluent of AP

Parameter	Point	Minimum	Maximum	Mean	Standard deviation
COD (mg.l⁻¹)	Influent	680	840	733	58
	Effluent	240	289	264	16
TSS (mg.l⁻¹)	Influent	100	195	152	32
	Effluent	28	93	47	21
pH (unit.)	Influent	6.6	7.2	-	-
	Effluent	6.5	6.9	-	-
Temperature (°C)	Influent	21.6	25.5	24.2	1.2
	Effluent	24.2	25.8	24.9	0.6

3.3.2 Sludge accumulation

The position of the inlets, points where GHG fluxes were measured, and outlets, as well as two-dimensional sludge profiles in the AP are shown in **Figure 3.1**. Solids accumulation was observed in the AP due to low maintenance. The last de-sludging at the time of this study was done five years before. The total sludge volume calculated in the AP was 2165 m³ (63% of total AP volume). This sludge accumulation occurred mainly in the last third of the lagoon and in the corners where the sludge layer reached up to 3.5 m. The deepest part of the lagoon was in the centre in the first half of the lagoon. Regarding the sludge accumulation rate this was in average 0.06 m³ per person per year, which was higher than the design value of 0.040 m³ per person per year suggested by Mara (1996). This suggests that sludge accumulation (due to lack of maintenance) has negatively affected treatment efficiency, leading to even higher sludge accumulation and lower biogas production.

3.3.3 Greenhouse gas fluxes

CO_2 and CH_4 fluxes

The CH_4 fluxes (**Figure 3.2**) measured in the three points of AP ranged from 13.37 to 178.7 L.m⁻².d⁻¹ (mean=76.1; s.d.= 48.2 n=31), while CO_2 ranged from 9.29 to 130.5 L.m².d¹ (mean=61.4;s.d.=32.2; n=31). The higher gas fluxes for both CH_4 and CO_2 were observed in May and October, which coincided with the largest observed COD removal in AP (**Table 3.2**). Likewise, during April and September CO_2 emissions were 1.5 and 1.3 times higher than CH_4.

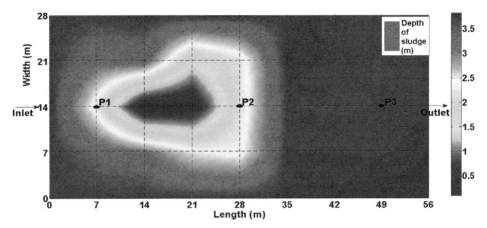

Figure 3.1 Sludge accumulation in the anaerobic pond. At the indicated points (P1, P2 and P3) GHG were measured.

Statistical analysis conducted to determine the correlation between the flux of gases produced and COD removed indicated that this variable explained the variation in fluxes of either CH_4 or CO_2 ($R_{Pearson} = 0.64$; $p<0.05$). No significant effect of temperature on the monthly variations of CH_4 and CO_2 flux (R Pearson=0.23) was observed. Likewise, there was a statistical correlation among CO_2 emissions and pH ($R_{Pearson}= 0.37$; $p<0.05$).

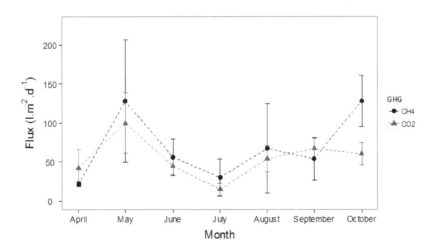

Figure 3.2 Profiles of CH_4 and CO_2 fluxes in AP.

Table 3.2 CH_4 and CO_2 conversion ratios in AP. Monthly data are average values from four measurements (n=28).

Month	COD removal (kg COD.d^{-1})	CH_4 production (m^3.d^{-1})	Ratio of CH_4 conversion (m^3.kgCOD$_{rem.}$$^{-1}$)	CO_2 production (m^3.d^{-1})	Ratio of CO_2 conversion (m^3.kgCOD$_{rem.}$$^{-1}$)
April	368.9	28.7	0.10	57	0.15
May	497.1	135.6	0.28	134.6	0.27
June	428.9	75.3	0.18	59.9	0.14
July	394.6	40.4	0.11	19.9	0.05
August	372.1	90.3	0.18	72.6	0.20
September	362.5	72.1	0.14	90.7	0.25
October	432.2	172.6	0.39	81.3	0.19

Based on CO_2 and CH_4 production and COD removal load, the conversion rate for these gases was calculated. The average amount of CH_4 was 0.24 ± 0.09 m^3.kgCOD$_{rem.}$$^{-1}$ and 0.18 ± 0.07 m^3 CO_2.kgCOD$_{rem.}$$^{-1}$ (**Table 3.2**). Based on the stoichiometry suggested by Metcalf *et al.* (2003) and under the temperature and atmospheric pressure conditions at Ginebra (T=26°C and b.p. = 0.89 atm.) these amounts were lower than the theoretically expected value of 0.43 m^3 CH_4.kgCOD$_{rem.}$$^{-1}$ and 0.35 m^3 CO_2.kgCOD$_{rem.}$$^{-1}$.

From a mass balance based on CDO influent to the AP it was found that 37% of the COD was transformed into biogas (CH_4 and CO_2, mainly) whereas 36% left the AP with the effluent. **Figure 3.3**. By assuming, a net biomass synthesis yield of 0.04 gVSS.gCOD$_{rem.}$$^{-1}$ (Metcalf and Eddy, 2003) the percentage of COD in the volatile solids produced was 3.5 %. Furthermore, the CH_4 dissolved in the effluent represents 2.5% while 21% of the influent COD was removed by other processes i.e. settling of solids to bottom.

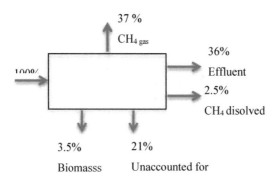

Figure 3.3 COD mass balance for the AP. Values show percentages of the influent COD.

Measurements at different points of the AP indicated a spatial variation of CH_4 and CO_2 emissions (**Fig. 3.4**). CH_4 emissions in the intermediate zone (136.6 ± 15.3 $l.m^{-2}.d^{-1}$) were almost twofold higher than those observed in the inlet zone (61.3 ± 10.8 $l.m^{-2} . d^{-1}$) and 3.5 times higher than those at the outlet (23 ± 5.6 $l.m^{-2}.d^{-1}$). Similar behaviour was observed for CO_2, in the intermediate zone CO_2 was 89.6 ± 39.4 $l.m^{-2}.d^{-1}$, which was 1.5 times higher than measured at the outlet (62.7 ± 34.6 $l.m^{-2}. d^{-1}$) and 2.2 times higher than measured in the inlet zone (41 ± 21.2 $l.m^{-2}.d^{-1}$).

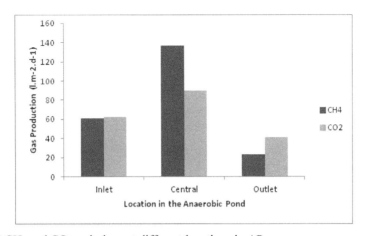

Figure 3.4 CH_4 and CO_2 emissions at different locations in AP

According to the measurements, the average biogas composition was $55 \pm 15\%$, and $32 \pm 6\%$ for CH_4 and CO_2 respectively. Furthermore, the comparison of CO_2 and CH_4 emissions for the three points showed that the composition of the biogas produced in the AP also showed a spatial variation (**Figure 3.4**). The gas concentrations at the inlet of the AP were 51.2% CH_4 and 39% for CO_2 while in the centre of the AP this was 67.4 % in CH_4 and 24% for CO_2. By contrast, in the area near to the output zone, the biogas produced was lower in CH_4 (34.6%) than CO_2 (58.1%).

N_2O fluxes
The N_2O fluxes measured in the AP were considerably lower than both CH_4 and CO_2 fluxes (**Figure 3.5**). In general, the median N_2O emissions for the months studied were 6.8 ± 3.6 $ml.m^{-2}.d^{-1}$. Although the highest N_2O emissions were observed in September and October, (12.7 and 13.1 $ml. m^{-2}.d^{-1}$, respectively) there were no significant differences between the different months monitored ($p < 0.05$). Further, there was no correlation between either pH or temperature and N_2O fluxes. A spatial variation in N_2O fluxes in the AP ($p < 0.05$), like for methane and nitrous oxide, was not found.

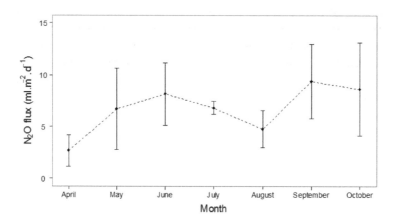

Figure 3.5 Profiles of N_2O flux in Ginebra`s AP

3.4 DISCUSSION

3.4.1 CH₄ and CO₂ fluxes

A comparison of the results obtained in this research and those reported in the literature can be seen in Table 3.3. As was expected the emissions of CO_2 and CH_4 from Ginebra AP were higher than those reported from full scale anaerobic ponds treating domestic wastewater under Mediterranean and subtropical climatic conditions (Toprak, 1995; Picot *et al.*, 2003; Hernandez-Paniagua *et al.*, 2014). Further, the average methane production of 76.1 $l.m^{-2}.d^{-1}$ obtained in this study was lower than the value of 97 $l.m^{-2}.d^{-1}$ measured in AP in Burkina Faso (Konaté *et al.*, 2013), which was operated under both temperatures and volumetric organic load higher than Ginebra's AP. On the other hand, the CH_4 emissions obtained in this study are lower than those reported for an AP operating with higher organic load and higher hydraulic retention time (DeSutter and Ham, 2005; Yacob *et al.*, 2005; Hasanudin *et al.*, 2006).

Based on methane produced and the COD removed, in the AP studied a yield of 0.24 m^3 CH_4/kg $COD_{removed}$ was calculated. This amounted to only 56% of the theoretical expected value and suggested as emission factor by IPCC (2006). Toprak (1995) reported an average value of 0.20 m^3 CH_4.kg $COD_{removed}^{-1}$ for an AP under temperate conditions, whereas Picot *et al.* (2003), measured in an AP under Mediterranean climate 0.36 m^3 CH_4.kg $COD_{removed}^{-1}$. Yacob *et al.* (2005) reported a mean conversion ratio of 0.34 m^3 CH_4/kg $COD_{removed}^{-1}$ for an AP treating palm oil mill effluent. All these different results in methane yield, including those obtained in this study, suggest that using only emission factors could underestimate or overestimate the methane emission figures. Therefore, local emission factors, resulting in a more realistic scenario which can be incorporated as part of national GHG inventories.

The average methane and carbon dioxide concentrations ($55 \pm 15\%$ CH_4; $32\pm6\%$ CO_2) found in this study, were in line with literature of anaerobic ponds treating palm oil mill effluent and tapioca wastewater (Yacob et al., 2005; Hasanudin et al., 2006). However, this situation was different to those reported for other anaerobic ponds, 52-80% CH_4 and 7-28% CO_2 in biogas from a full-scale anaerobic pond in Portugal (Toprak, 1995), 83% CH_4 and 4% CO_2 in an anaerobic pond under Mediterranean climatic conditions (Picot et al., 2003), and 80% CH_4 and 2.5 %CO_2 in biogas from an AP under the Sudano-Sahelian climate of Burkina Faso (Konaté et al., 2013). These differences may be explained by the slightly alkaline conditions (pH=7.8), since a pH increase will result in a higher CO_2 conversion to bicarbonate alkalinity in the water column decreasing the amount of CO_2 released into the biogas (Green et al,. 1995).

Table 3.3 Emissions of greenhouse gases from anaerobic ponds as reported in literature

Wastewater source	Climate	CH_4	CO_2	Reference
		l CH_4.m^{-2}.d^{-1}	l CO_2.m^{-2}.d^{-1}	
Municipal	Tropical	13.37 - 178.7 (76.1)	9.29- 130.5 (61.4)	This study
Municipal	Mediterranean	18.2-48.7 (31.6)	3.1-9.8 (5.5)	Toprak (1995)
Municipal	Mediterranean	10.79-89-64(40.7)	-	Picot et al. (2003)
Livestock facilities	Temperate	92.3-323 (n.d.)	-	DeSutter and Ham (2005)
Palm oil effluent	Tropical	391.7-1880 (1228)	-	Yacob et al. (2005)
Tapioca WW	Temperate	516.4-993(n.d.)	-	Hasanudin et al. (2006)
Municipal	Temperate	3.3-5.28	-	Wang et al. (2011)
Municipal	Sahelian	28-189 (97)	-	Konaté et al. (2013)
Agricultural WWT	Subtropical	10.8±1.5	0.6±0.4	Hernandez-Paniagua et al. (2014)
Municipal	Subtropical	42.7±15.9	-	Paredes et al. (2015)

Data in parenthesis represent average measurements

Organic matter removal and temperature have been identified as primary variables influencing the generation of GHG i.e. CH_4 and CO_2 from APs (Toprak, 1995; Picot et al., 2003; DeSutter and Ham, 2005; Yacob et al., 2005; Konaté et al., 2013). In this study, a positive correlation among changes in the flux of CO_2 and CH_4 and COD removal was found. A low influent COD leads to lower GHG production i.e. CO_2 and CH_4 (**Figure 3.3**). No statistical correlation between temperature and GHG emissions was found. A possible reason was the minor variation in temperature in the water column which impacted the sensitivity of the statistical correlation test (temperature was almost constant during the measurement period: 24.2 ± 1.2 °C). Despite this, the emissions measured in this study, under tropical conditions, were higher than observed in other studies at temperatures below 20 ° C (**Table 3.3**).

According to the mass balance, the COD conversion to methane was only 37%. This value was lower than reported in anaerobic ponds for the primary treatment of urban wastewater treatment in a Mediterranean climate (Picot *et al.*, 2003), but higher than estimated in a subtropical lagoon (Hernandez-Paniagua *et al.*, 2014). This result could be influenced by the measurement method used to estimate methane flux. Methane from anaerobic pond can be released into the atmosphere by ebullition and diffusion. Ebullition is related to large, spontaneous eruptions of methane accumulated in the sludge layer. Diffusion is due to concentrations differences between air-water interphases. The static chamber is designed to measure methane release by diffusion. Thus, a part of the methane emitted by ebullition was not accounted for, giving an under-estimation of the actual daily emission.

In addition, based on the results obtained from the mass balance, it would seem that both COD accumulation in the sludge, and CH_4 dissolved in the effluent have resulted in a lower methane yield which, which may also affect the subsequent treatment stages. Sludge accumulation may lead to the washout of re-suspended solids containing biodegradable organic matter into a secondary facultative pond i.e. algae facultative pond (AFP). Such peak loads in AFP may result in anoxic situations, leading to the release of CH_4, CO_2 and N_2O. Furthermore, the CH_4 dissolved in the effluent of the AP that is conducted into AFP probably, will probably be transferred into the atmosphere by both ebullition and diffusion processes. Therefore, to prevent GHG emissions in subsequent treatment systems the AP should be designed to optimize COD removal and for the capture and utilization of methane.

In this study, a spatial variation of CH_4 and CO_2 emissions from the AP were observed. The non-uniform sludge accumulation in different zones of the pond could explain this variation (**Figure 3.2**). Probably the high sludge accumulation in the inlet and outlet locations of Ginebra`s AP changed its hydraulic behaviour and thus the biochemical transformation of organic matter into biogas was influenced. The hydraulic efficiency of ponds is compromised by sludge accumulation, which affects pond performance by reducing effective volume decreasing hydraulic residence times(Peña *et al.*, 2000; Keffala *et al.*, 2012). In particular, this will affect the quality of treated water, including, but not limited to, higher biological oxygen demand (BOD), total suspended solids, nitrogen and phosphorus (N and P), and less pathogen removal. Likewise, less organic matter is converted into methane and carbon dioxide.

Papadopoulos *et al.* (2003) distinguished three different zones in the sludge layer of an AP. The first zone, one of high-density sludge lying at the bottom of the pond, consisted of inert solid sludge. The second zone, lying above this, was characterized by a high concentration of volatile (easily biodegradable) sludge. The third zone (supernatant), lying at the top of the pond, was identified as a liquid layer, low in suspended solids. According to this description, in the point P2, probably the sludge layer was higher in biodegradable (volatile) organic matter and biogas production was higher than in the P1 and P3. By contrast, the outlet zone was dominated by sludge accumulation of high-density (inorganic), which is not easily biodegradable and therefore the biogas production was limited. These results highlight the importance to implement best practices of implementing O&M best practice, in order to decrease the probability of excessive sludge accumulation and malfunctioning of AP.

3.4.2 N₂O fluxes

Considering that N_2O has a high global warming potential, studies of GHG emissions from wastewater treatment systems should include systematic measurements of this gas. This presents a gap in literature, as previous studies have not systematically looked at N_2O emissions from AP treating municipal wastewater. This may have been due to the assumption that anoxic systems will be unlikely to produce N_2O. However, in this study was found a positive N_2O flux of 6.8 ± 3.6 ml.m^{-2}.d^{-1} (12.2 ± 6.4 mg.m^{-2}.d^{-1}) indicating that in AP this gas could be produced.

N_2O is generated in wastewater treatment systems with high organic load and low oxygen concentrations (Law *et al.*, 2012). However, the N_2O production in the AP is constrained by the oxidation of ammonium to nitrate (NO_3^--N), which requires an appropriate oxygen concentration level. One hypothesis is that the Ginebra`s AP had a relatively large surface area and possibly some oxygen was transferred via diffusion to the water surface favouring some NH_4^+-N oxidation and producing NO_3^--N. This NO_3^--N was promptly consumed by denitrification producing N_2O. Given that in this study, the extension under which oxygen was transferred into AP and the NO_3^--N consumption were not determined, it is difficult to make conclusions on the main processes responsible for N_2O production in the AP studied.

The Intergovernmental Panel on Climate change (IPCC, 2006) proposes an emission factor of 0.35 gN₂O-N.(kg TKN$_{influent}$)$^{-1}$ for developed countries, characterized by a high protein intake. When this factor is calculated to AP studied it was obtained a value of 0.68 gN₂O-N.(kg TKN$_{influent}$)$^{-1}$, i.e. 0.068% of the incoming nitrogen. This suggest that using emission factor there was an underestimation around two-folds regarding to field measurements of N_2O emissions produced in the anaerobic pond. In conventional WWT systems also have been reported differences between the value of N_2O measured in situ and value recommended by IPCC (2006). In these researches were reported values ranging from 0.001 to 14.6% of the incoming nitrogen (Kampschreur *et al.*, 2009; Ahn *et al.*, 2010; Foley *et al.*, 2015). This demonstrates again the inconveniences of using a single emission factors to estimate GHG from wastewater treatment. Therefore, more measurements are necessary to provide quantitative GHG emission data for various wastewater treatment systems.

3.4.3 Greenhouse Gas emissions in terms of CO₂ equivalent

For the conversion of measured greenhouse gases into actual global warming potential (GWP), the IPCC suggests only taking CH₄ and N₂O into consideration with 28 and 265 CO₂ equivalents (IPCC, 2007). Based on the GHG emissions obtained in the current study, the N_2O amount was 0.003 kg CO$_{2eq}$.m^{-2}.d^{-1} and CH₄ around 1.39 kg CO$_{2eq}$.m^{-2}.d^{-1}. Thus the total CO₂ equivalent for Ginebra′s AP was around 1.4 kg CO$_{2eq}$.m^{-2}.d^{-1}. Taking into account only the emissions of CH₄ (1.39 kgCO$_{2eq}$.m^{-2}.d^{-1}), the dimensions of the AP (714 m^2 and 9000 PE), and

COD removed (405.2 $kgCOD_{removed}.d^{-1}$), suggests a GWP total for the AP of 37.8 $kgCO_{2eq}.PE^{-1}.y^{-1}$ or 2.18 $kgCO_{2eq}.kgCOD_{removed}^{-1}$. By comparison, in conventional aerobic/anaerobic full treatment systems, values ranging between 0.91 and 1.04 kg $CO_{2eq}.kgCOD_{removed}^{-1}$ (Keller and Hartley, 2003) and 1.0 to 1.4 kg $CO_{2eq}.kgCOD_{removed}^{-1}$ (Flores-Alsina et al., 2011) were reported.

According to these figures and taking into account that AP is only a wastewater pre-treatment, it is expected that eco-technologies based on full stabilization ponds have a higher global warming impact than conventional systems i.e. activated sludge. However, this high footprint can be reduced depending on stabilization pond design, e.g. by including capture and use of biogas

Based on the figures obtained for Ginebra's AP and assuming that in Colombia only 9% of all wastewater is treated and 55% of it through WSPs (IDEAM, 2008) it can be estimated that a total of 1807 $tonCO_{2eq}.d^{-1}$ (0.66$MtCO_{2eq}yr^{-1}$) could be produced from APs in Colombia. This is around 0.3% of the total GHG emissions reported for Colombia (180 $MtCO_{2eq}.yr^{-1}$). However, the emissions from wastewater treatment in Colombia would be higher if also the 91% of non-treated wastewater were taken into account. These untreated effluents would probably generate similar amounts of CO_{2eq} after their discharge into aquatic environment. Therefore, the total contribution of wastewater effluents (treated and untreated) to overall GHG emissions in Colombia is estimated at 3.2% (IDEAM, 2008).

3.5 CONCLUSIONS

- This study provided information about greenhouse gas emissions i.e. CO_2, CH_4 and N_2O from an anaerobic pond treating municipal wastewater. The production of these gases was influenced by tropical conditions and operation conditions i.e. temperature, organic load and sludge accumulation. In particular, sludge accumulation could change the hydraulic of the AP and COD removal was low affecting the biogas quality.

- This study confirms that using single emission factors to estimate greenhouse gas emission from wastewater treatment could generate uncertainties (under or over estimations).

- In this study N_2O positive fluxes were observed in the AP. Although it was not possible to completely explain the mechanism leading to N_2O production, this highlights the necessity of more studies involving this topic.

- The Ginebra's AP showed a larger greenhouse gas footprint compared to conventional systems i.e. activated sludge. This could be changed if effective capture and utilization of the biogas produced was included in the design of the AP. When an AP is operated as an open system (non-capture) all methane is released into the atmosphere and therefore the AP will have a larger impact on the GHG phenomenon compared to conventional systems.

3.6 ACKNOWLEDGEMENTS

This study has been carried out within the framework of the European research project SWITCH (Sustainable Urban Water Management Improves Tomorrow's City's Health). SWITCH is supported by the European Commission under the 6th Framework Programme and contributes to the thematic priority area of "Global Change and Ecosystems" [1.1.6.3] Contract n° 018530-2. The authors are also indebted to ACUAVALLE ESP and the Universidad del Valle for their support.

3.7. REFERENCES

Ahn, J.H., Kim, S., Park, H., Rahm, B., Pagilla, K. and Chandran, K. (2010). N_2O emissions from activated sludge processes, 2008– 2009: results of a national monitoring survey in the United States. Environmental Science & Technology 44(12), 4505-4511.

APHA (2005) APHA,AWWA, WEF, Standard methods for the examination of water and wastewater, American Public Health Association, American Water Works Association and Water Environment Federation, 21 st ed. Washington DC.

Bogner, J., M. Abdelrafie Ahmed, C. Diaz, A. Faaij, Q. Gao, S. Hashimoto, K. Mareckova, R. Pipatti and Zhang, T. (2007). Waste Management, In Climate Change 2007: Mitigation. Contribution of Working Group III to the Fourth Assessment Report of the Intergovernmental Panel on Climate Change. B. Metz, O.R. Davidson, P.R. Bosch, R. Dave and Meyer, L.A. (eds), p. 32, Cambridge University, Cambridge, United Kingdom

Cakir, F.Y. and Stenstrom, M.K. (2005). Greenhouse gas production: a comparison between aerobic and anaerobic wastewater treatment technology. Water Research 39(17), 4197-4203.

DeGarie, C.J., Crapper, T., Howe, B.M., Burke, B.F. and McCarthy, P.J. (2000). Floating geomembrane covers for odour control and biogas collection and utilization in municipal lagoons. Water Science & Technology 42(10), 291-298.

DeSutter, T.M. and Ham, J.M. (2005). Lagoon-biogas emissions and carbon balance estimates of a swine production facility. Journal of environmental quality 34(1), 198-206.

El-Fadel, M. and Massoud, M. (2001). Methane emissions from wastewater management. Environmental Pollution 114(2), 177-185.

EPA, U. (2012). Global Anthropogenic Non-CO2 Greenhouse Gas Emissions: 1990–2030. Office of Atmospheric Programs, Climate Change Division. WashingtonD.C: US Environmental Protection Agency, US EPA Washington, DC.

Flores-Alsina, X., Corominas, L., Snip, L. and Vanrolleghem, P.A. (2011). Including greenhouse gas emissions during benchmarking of wastewater treatment plant control strategies. Water Research 45(16), 4700-4710.

Foley, J., Yuan, Z., Keller, J., Senante, E., Chandran, K., Willis, J., Shah, A., van Loosdrecht, M.C. and van Voorthuizen, E. (2015) N_2O and CH_4 Emission from Wastewater Collection and Treatment Systems: State of the Science Report and Technical Report, IWA Publishing.

Green, F.B., Bernstone, L., Lundquist, T.J., Muir, J., Tresan, R.B. and Oswald, W.J. (1995). Methane fermentation, submerged gas collection, and the fate of carbon in advanced integrated wastewater pond systems. Water Science & Technology 31(12), 55-65.

Hasanudin, U., Utomo, T., Suroso, E., Hendri, J., Inokawa, A. and Fujie, K. (2006). Methane and CO2 Gases Emission from Anaerobic Pond in Tapioca Wastewater Treatment, Bangkok, Thailand.

Hernandez-Paniagua, I., Ramirez-Vargas, R., Ramos-Gomez, M., Dendooven, L., Avelar-Gonzalez, F. and Thalasso, F. (2014). Greenhouse gas emissions from stabilization ponds in subtropical climate. Environmental technology 35(6), 727-734.

IDEAM (2008). Segunda Comunicación Nacional de Colombia ante la Convención Marco de las Naciones Unidas sobre Cambio Climático. IDEAM (ed), IDEAM, Bogotá.

IPCC (2006). 2006 IPCC Guidelines for national greenhouse gas inventories: waste. Eggleston, S., Buendia, L. and Miwa, K. (eds), Institute for Global Environmental Strategies, Kanagawa, Japan.

IPCC (2007). Fourth Assessment Report of the Intergovernmental Panel on Climate Change, Cambridge UCambridge University Press. United Kingdom and New York, NY, USAniversity Press. United Kingdom and New York, NY, USA.

Kampschreur, M.J., Temmink, H., Kleerebezem, R., Jetten, M.S. and van Loosdrecht, M.C. (2009). Nitrous oxide emission during wastewater treatment. Water Research 43(17), 4093-4103.

Karakurt, I., Aydin, G. and Aydiner, K. (2012). Sources and mitigation of methane emissions by sectors: A critical review. Renewable Energy 39(1), 40-48.

Keffala, C., Harerimana, C. and Vasel, J.-l. (2012). A review of the sustainable value and disposal techniques, wastewater stabilisation ponds sludge characteristics and accumulation. Environmental Monitoring and Assessment, 1-14.

Keller, J. and Hartley, K. (2003). Greenhouse gas production in wastewater treatment: process selection is the major factor. Water Science & Technology 47(12), 43-48.

Konaté, Y., Maiga, A.H., Casellas, C. and Picot, B. (2013). Biogas production from an anaerobic pond treating domestic wastewater in Burkina Faso. Desalination and Water Treatment 51(10-12), 2445-2452.

Law, Y., Ye, L., Pan, Y. and Yuan, Z. (2012). Nitrous oxide emissions from wastewater treatment processes. Philosophical Transactions of the Royal Society B: Biological Sciences 367(1593), 1265-1277.

Lettinga, G. (1995). Anaerobic digestion and wastewater treatment systems. Antonie van leeuwenhoek 67(1), 3-28.

Lettinga, G., Rebac, S. and Zeeman, G. (2001). Challenge of psychrophilic anaerobic wastewater treatment. TRENDS in Biotechnology 19(9), 363-370.

Mara, D. (1996) Low-cost urban sanitation, John Wiley & Sons, Cichester, England.

Mara, D.D. (2004) Domestic wastewater treatment in developing countries, Earthscan, London.

Mara, D.D. and Pearson, H.W. (1998) Design manual for waste stabilization ponds in Mediterranean countries, Lagoon Technology International Ltda.

Metcalf, L. and Eddy, H. (2003) Wastewater Engineering: Treatment, Disposal and Reuse, , Mc Graw Hill, New York.

Myhre, G., Shindell, D., Bréon, F.-M., Collins, W., Fuglestvedt, J., Huang, J., Koch, D., Lamarque, J.-F., Lee, D. and Mendoza, B. (2013). Anthropogenic and natural radiative forcing. Climate change 423.

Papadopoulos, A., Parisopoulos, G., Papadopoulos, F. and Karteris, A. (2003). Sludge accumulation pattern in an anaerobic pond under Mediterranean climatic conditions. Water Research 37(3), 634-644.

Paredes, M., Güereca, L., Molina, L. and Noyola, A. (2015). Methane emissions from stabilization ponds for municipal wastewater treatment in Mexico. Journal of Integrative Environmental Sciences, 1-15.

Peña, M.R., Rodriguez, J., Mara, D.D. and Sepulveda, M. (2000). UASBs or anaerobic ponds in warm climates? A preliminary answer from Colombia. Water Science & Technology 42(10), 59-65.

Picot, B., Paing, J., Sambuco, J.P., Costa, R.H.R. and Rambaud, A. (2003). Biogas production, sludge accumulation and mass balance of carbon in anaerobic ponds. Water science and technology 48(2), 243-250.

Shahabadi, M.B., Yerushalmi, L. and Haghighat, F. (2009). Impact of process design on greenhouse gas (GHG) generation by wastewater treatment plants. Water Research 43(10), 2679-2687.

Shilton, A.N., Powell, N., Mara, D.D. and Craggs, R. (2008). Solar-powered aeration and disinfection, anaerobic co-digestion, biological CO_2 scrubbing and biofuel production: the energy and carbon management opportunities of waste stabilisation ponds. Water science and technology 58(1), 253-258.

Silva, J.P., Lasso, A., Lubberding, H.J., Peña, M.R. and Gijzen, H.J. (2015). Biases in greenhouse gases static chambers measurements in stabilization ponds: Comparison of flux estimation using linear and non-linear models. Atmospheric Environment 109, 130-138.

Sweetapple, C., Fu, G. and Butler, D. (2014). Identifying sensitive sources and key control handles for the reduction of greenhouse gas emissions from wastewater treatment. Water Research 62, 249-259.

Toprak, H. (1995). Temperature and organic loading dependency of methane and carbon dioxide emission rates of a full-scale anaerobic waste stabilization pond. Water Research 29(4), 1111-1119.

Wang, J., Zhang, J., Xie, H., Qi, P., Ren, Y. and Hu, Z. (2011). Methane emissions from a full-scale A/A/O wastewater treatment plant. Bioresource technology 102(9), 5479-5485.

Yacob, S., Hassan, M.A., Shirai, Y., Wakisaka, M. and Subash, S. (2005). Baseline study of methane emission from open digesting tanks of palm oil mill effluent treatment. Chemosphere 59(11), 1575-1581.

Chapter 4

Influence of the photoperiod on carbon dioxide, methane and nitrous oxide emissions from two pilot-scale stabilization ponds

This Chapter has been presented and published as:

Silva J.P., Ruiz JL, Lubberding HJ, Peña MR, and Gijzen HJ (2012). Influence of photoperiod on carbon dioxide and methane emissions from two pilot-scale stabilization ponds. Water Science and Technology, 66, 1930-1940.

Silva J.P., Ruiz JL, Lubberding HJ, Peña MR, and Gijzen HJ (2011). Influence of photoperiod on carbon dioxide and methane emissions from two pilot-scale stabilization ponds in: 9th IWA stabilization ponds, August 1-3, 2011, Adelaide, AU.

Abstract

Greenhouse gas emissions (CO_2, CH_4, and N_2O) from pilot-scale algae facultative and duckweed-based ponds (AFP and DBP) were measured using the static chamber methodology. Daytime and night-time variations of GHG and wastewater characteristics e.g. COD, and pH were determined via sampling campaigns during the midday (12:30-15:30) and middle of the night (00:30-03:30). The results showed that under daytime conditions in the AFP median emissions were -232 mg CO_2 m^{-2} d^{-1}, 9.9 mg CH_4 m^{-2} d^{-1}, and 6.9 mg N_2O m^{-2} d^{-1}, and in the DBP median emissions were -1,654.5 mg CO_2 m^{-2} d^{-1} and 71.4 mg CH_4 m^{-2} d^{-1}, and 8.5 mg N_2O m^{-2} d^{-1} respectively. During night-time conditions the AFP median emissions were 3,949.9 mg CO_2 m^{-2} d^{-1}, 12.7 mg CH_4 m^{-2} d^{-1}, and 5.5 mg N_2O m^{-2} d^{-1} whereas the DBP median emissions were 5,116 mg CO_2 m^{-2} d^{-1}, 195.2 mg CH_4 m^{-2} d^{-1}, and 2 mg N_2O m^{-2} d^{-1} respectively. Once data measured during the daytime were averaged together with night-time data the median emissions for the AFP were 1,566.8 mg CO_2 m^{-2} d^{-1}, 72.1 mg CH_4 m^{-2} d^{-1}, and 9.5 mg $N_2O.m^{-2}.d^{-1}$ whilst for the DBP they were 3,016.9 mg CO_2 m^{-2} d^{-1}, 178.9 mg CH_4 m^{-2} d^{-1}, and 8.6 mg $N_2O.m^{-2}.d^{-1}$. These figures suggest that there were significant differences between CO_2 emissions measured during daytime and night-time periods ($p<0.05$) signifying a sink-like behaviour for both the AFP and DBP in the presence of solar light, which indicates the influence of photosynthesis in the CO_2 emissions. Overall, according to the compound average (daytime and night-time) both AFP and DBP systems might be considered as net sources of GHG.

4.1 INTRODUCTION

The atmospheric concentrations of greenhouse gases (GHG) such as carbon dioxide (CO_2), methane (CH_4), and nitrous oxide (N_2O) have continuously risen since the pre-industrial age. According to Forster et al. (2007), a wide range of direct and indirect measurements confirm that the atmospheric mixing ratios of CO_2, CH_4, and N_2O have increased globally over the last 250 years by 36%, 250%, and 18%, respectively. Human activities - e.g. electrical power production, industrial processes, agriculture, forestry, and waste management (solid waste and wastewater) - are by far the major contributors to the GHG increase. Thus, compilation of data covering these sectors is the basis for collective action on the reduction of anthropogenic GHG emissions (UNFCCC, 2007).

GHG emissions from wastewater treatment (WWT) are estimated to represent less than 5% of the total emission load (Bogner et al., 2007). However, these figures exhibit uncertainties because wastewater data are missing, inconsistent or incomplete, especially in developing countries. In addition, the use of the emission factors method (IPCC, 2006) to estimate GHG emissions from WWT would be the major source of uncertainty for many developing countries, because these default or theoretical values do not take into account the different processes involved in WWT technologies (El-Fadel and Massoud, 2002). It has to be said that any measure that penalizes emission of GHG (e.g. via carbon or GHG taxes) or imposes mandatory limitations on their release will also impact the operation of WWT plants (Greenfield *et al.*, 2005); therefore, measuring actual GHG from WWT is considered of great importance.

Waste stabilization ponds (WSPs) are efficient, low-cost and low-tech options for sewage treatment mainly in developing countries (Arthur, 1983; Peña *et al.*, 2002; Mara, 2005). WSPs use little or no electrical energy, are more appropriate than energy-intensive processes, such as activated sludge, and they are cheaper to construct, operate and maintain. However, WSPs may generate secondary negative environmental impacts because they may generate greenhouse gases such as carbon dioxide (CO_2), methane (CH_4) and nitrous oxide (N_2O) related to the intrinsic metabolic processes that occur whilst achieving the desired degree of treatment (Van der Steen *et al.*, 2003b).

Wastewater treatment by two types of WSPs, AFP and DBP, has been studied for the removal of carbon, nitrogen, bacteria and viruses (Toprak, 1995; Nalbur *et al.*, 2003; Awuah *et al.*, 2004; Zimmo *et al.*, 2004a; El-Shafai *et al.*, 2007; Johnson and Mara, 2007). However, only a few studies have considered GHG emissions from AFPs and DBPs (Van der Steen *et al.*, 2003a; Singh *et al.*, 2005; Stadmark and Leonardson, 2005). Besides, these studies were carried out in temperate regions where environmental factors such as solar radiation and temperatures are quite different from those in tropical regions.

Hence, this study was aimed at measuring the emissions of CO_2, CH_4, and N_2O from a pilot-scale AFP and DBP. In addition, the influences of environmental factors that regulate the emissions were also evaluated. Special emphasis was given to estimating differences in emissions during daytime (higher solar radiation) and night-time (lower solar radiation) conditions, because in tropical settings other factors (e.g. photosynthetic activity) might have a stronger effect on GHG emissions than the wastewater composition. Finally, this research was also aimed at contributing to a reduction in the knowledge gap on GHG emissions field data from AFPs and DBPs.

4.2 MATERIALS AND METHODS

4.2.1 Experimental set-up
This study was carried out on pilot-scale units located in the experimental research station for wastewater treatment and reuse at Ginebra, a small town of 10,000 inhabitants located in south-west Colombia (3°43'25.98 N, 76°15'59.45 W). The historical average ambient temperature of this town is 26°C and its altitude is 1,040 m above sea level. The pilot plant consisted of two parallel systems: a DBP system seeded with *Spirodela polyrrhiza* and an AFP system. The latter was essentially a pilot facultative pond. Each system consisted of three identical plastic cylindrical tanks (0.40 m in water depth and 0.56 m in diameter) in series operating with a theoretical hydraulic retention time of 4 days (**Figure 4.1**). Both systems received the effluent

from an up-flow anaerobic sludge blanket (UASB) reactor with an average hydraulic retention time of 7-8 hours.

The operational parameters for both systems are shown in **Table 4.1**. The AFP and DBP were started operating simultaneously at a constant influent flow. After 1 month of stabilization (i.e. until steady state conditions were reached) a sampling campaign was undertaken over a period of 6 months for GHG and wastewater quality measurements under different solar radiation conditions: from 12:30 to 15:30 (daytime period), which is the period of maximum solar radiation in Ginebra, and from 00:00 to 03:00 (night-time period), which is the period of minimum solar radiation (darkness).

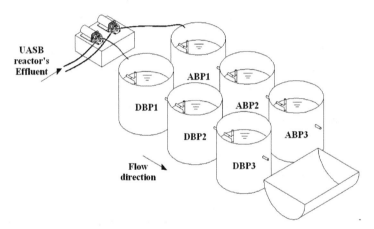

Figure 4.1 Experimental set-up of the pilot plant. DBP= duckweed-based ponds; AFP= algae facultative pond

Table 4.1 Operational characteristics of the AFP and DBP pilot ponds

Parameter	Units	DBP	AFP
Flow	l.d^{-1}	24.5	24.5
Depth	m	0.4	0.4
HRT per pond	d	4	4
Total HRT	d	12	12
Hydraulic surface loading	m^3·m^2·d^{-1}	0.09	0.09
Organic loading for the first pond	kg BOD$_5$ ha^{-1}d^{-1}	77.7* – 139.4**	77.7*– 139.4**
Biomass density (after harvesting)	g.m^{-2}	700-1,000	Not applicable

* Daytime period, ** Night-time period

4.2.2 Wastewater sampling

Biochemical oxygen demand (BOD_5), chemical oxygen demand (COD), total suspended solids (TSS), alkalinity, total Kjeldahl nitrogen (TKN), ammonium nitrogen (NH_4^+-N), and nitrate nitrogen (NO_3^--N) were measured according to Standard Methods (APHA *et al.*, 2005). Conductivity, pH, dissolved oxygen, temperature and oxidation-reduction potential were measured with electrodes.

The biomass density in the DBP was assessed by gravimetry, determining the amount to be harvested so as to leave a density of 700-1,000 $g.m^{-2}$ (fresh weight) (Caicedo, 2005). In the AFP chlorophyl-a was measured using an ultraviolet light *Aquaflor* (Turner Designs) handheld fluorometer.

4.2.3 Greenhouse gas measurements

Greenhouse gas emissions (CH_4 and CO_2, CH_4, and N_2O) were determined by using the static chamber technique. All the chambers were constructed in acrylic Plexiglas (0.3 m x 0.3 m: diameter x height) and supplied with a thermometer and a sampling port. The chambers were fixed at the water surface in the central part of the ponds. Samples of the gas (20 ml) for CH_4 and CO_2 measurements were taken over a period of 30 minutes at 10-minute intervals (0, 10, 20, 30 min) from a sampling port on the top of each chamber using a special syringe. Finally, the gas samples were withdrawn directly through a needle into pre-evacuated containers with a volume of 10 ml.

CH_4 was analysed by gas chromatography (Shimadzu Co., Japan) with a flame ionic detector (FID) and a Porapak Q column, and the temperature of the oven and injector ports were set at 60°C and 80°C, respectively. N_2O concentration was analysed by means of gas chromatography (Shimadzu Co., Japan) equipped with an electron capture detector (ECD) and a Porapak column Q 80-100 mesh 2 m*2 mm retention gap, using 22 ml min^{-1} N_2 was the carrier gas, and the temperature at the injector, column, and detector were 80, 70, and 320°C, respectively. The flow rate of the carrier gas (N_2) was 22 ml min^{-1}. CO_2 was measured by an infrared spectrophotometer (Qubit S151 CO_2 analyser) using 75 mL min^{-1} air as the mobile phase; the temperature of the injector was set equal to the ambient temperature.

4.2.4 Flux calculation

In this paper, the term flux is defined as gas transfer per surface area ($ML^{-2}T^{-1}$) and the term emission is used for positive fluxes, while the term consumption is used for negative fluxes. CH_4 and CO_2 concentrations obtained from chromatographic and infrared analysis were processed in a spreadsheet to calculate net fluxes. Gas fluxes were calculated from linear and non-linear changes in the gas concentration in the chamber headspace according to the protocol described by Silva *et al.* (2015). Flux figures were accepted only when the coefficient of correlation (R^2) was equal or higher than 0.85 as recommended by other studies (Huttunen *et al.*, 2002; Lambert and Fréchette, 2005).

4.2.5 Data analysis
Statistical analyses were done with SPSS® software (v. 15.0 for Windows). A nonparametric statistical test (i.e. Mann-Whitney U-test) was applied to test whether the behaviour of both the AFP and DBP in terms of final water quality and GHG emissions was significantly different at the significance level $\alpha = 0.05$.

4.3 RESULTS

4.3.1 Performance of the AFP and DBP
The characteristics of the influent and effluent of the AFP and DBP are shown in **Table 4.2**. The influent was almost neutral (pH 6.6-6.8) and anaerobic, because it was pre-treated in a UASB reactor and thus, it had a negative oxidation-reduction potential (ORP) and high NH_4^+-N concentrations.

Table 4.2 Wastewater characteristics (mean $\pm$ SD) for influent and effluent of an algae-based pond (AFP) and a duckweed-based pond (DBP)

Parameter	Influent	AFP effluent	DBP effluent
$COD^£$ (mg.l⁻¹)	124 ± 35.9	47.1 ± 10.5	42.2 ± 8
TKN (mg.l⁻¹)	38.3 ± 6.9	9.5 ± 4.7	20.9 ± 3.6
NH_4^+-N (mg.l⁻¹)	35.7 ± 4.4	1.0 ± 0.8	19.3 ± 3.6
NO_3^--N (mg.l⁻¹)	2.6 ± 2.6	5.2 ± 1.9	2.3 ± 1.8
pH** (Un)	6.8 ± 0.3	7.7 ± 0.3	7.4 ± 0.2
ORP (mV)	-286 ± 49	152.7 ± 39.1	127.8 ± 61.5

*Calculated based on COD soluble (n=15)

**Calculated based on average $[H_3O^+]$ values obtained during daytime and night-time

No significant differences were found between the effluents of the AFP and DBP in relation to soluble COD ($p>0.05$). In general, the DBP provided a better quality of effluent in terms of organic matter content (lower concentrations of soluble COD), whereas the AFP provided a better effluent than the DBP in terms of nitrogen content (lower concentrations of TKN and NH_4^+-N). In addition, NO_3^--N concentrations in both the AFP and DBP were higher than those found in the influent, indicating nitrification; nitrates were regularly higher in the AFP effluent, a fact that suggests stronger oxidation conditions in the AFP compared to the DBP.

The effluent pH values found in the DBP and AFP during the daytime period ranged from 6.9 to 7.9 and 7.0 to 9.2, respectively. On the other hand, under nocturnal conditions the pH variations were similar (7.3-7.6) for both the DBP and AFP. The high pH values observed in the AFP can be explained by the photosynthetic activity of suspended algae, which were much less dominant in the duckweed ponds due to the presence of a duckweed cover that provided shading of the water column.

4.3.2 Greenhouse gas emissions
CO₂ fluxes

The median daytime CO_2 fluxes (**Figure 4.2**) in the AFP ranged from -2,963 to 6,403 mg.m^{-2} .d^{-1} (Median = -232; SD = 2,682; n = 23) while in the DBP it ranged from -5,039.2 to 4,439 mgm^{-2}d^{-1} (Median = -1,654.5; SD = 2,673.6; n = 24). In general, the CO_2 rates measured in the AFP and DBP ponds suggest that both emission and consumption occurred.

A high percentage (>60%) of measurements showed negative fluxes that indicated a sink-like (consumption) behaviour for both the AFP and DBP during daylight (**Figure 4.2**). In addition, according to the statistical analysis there were no significant differences in daytime fluxes (p = 0.053) emitted from the AFP and DBP, and thus these ponds showed a similar sink potential for CO_2. On the other hand, during the night-time period both the AFP and DBP showed only CO_2 positive fluxes and the AFP ranged from 1,061 to 7,655 mg.m^{-2}.day^{-1} (Median = 3,949.9; SD = 2,066.4; n = 17), and DBP ranged from 3,101.1 to 9,898 mg.m^{-2}.day^{-1} (Median = 5,116.3; SD = 1,901.1; n = 18) (**Figure 4.2**). Overall, during daytime both the AFP and DBP showed CO_2 sink, while under night-time conditions both the AFP and DBP showed a CO_2 source behaviour. Finally, when data measured during daylight were averaged together with nocturnal data, the 24-hour median emissions for the AFP and DBP were 1,566.8 mg CO_2 m^{-2} d^{-1} and 3,016.9 mg CO_2 m^{-2} d^{-1} respectively.

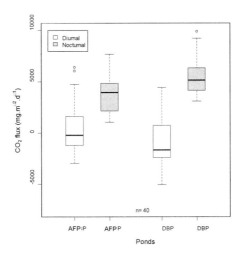

Figure 4.2 Fluxes of CO_2 from the AFP and DBP measured during daytime and night-time periods. In the whiskers are showed the quantile: Q1, Q2, and Q3; Q2 = Median; IRQ = Q3-Q1; Min = Q1-1.5IRQ; Max = Q3 + 1.5IRQ.

In **Figure 4.3a** can be observed the CO_2 fluxes measured in each cylindrical reactor of the AFP and DBP for the daytime period. During the passage of the wastewater through the three successive cylindrical reactors the CO_2 fluxes in the DBP ponds increased along the series, whilst a decreasing trend was observed in the AFP. Likewise, the daytime CO_2 emissions in the first tank were significantly higher for the AFP than for the DBP ($p<0.05$), while differences observed in CO_2 emissions for the second and third reactor of both ponds were not significant ($p = 0.41$).

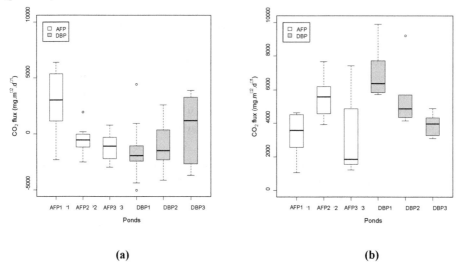

(a)   (b)

Figure 4.3 Fluxes of CO_2 from the AFP and DBP measured in each cylindrical reactor during daytime (a) and night-time (b) periods

In addition, for both the AFP and DBP the nocturnal CO_2 fluxes were positive (**Figure 4.3b**), which indicates the ponds were a net source of CO_2. A decreasing trend in CO_2 fluxes from the DBP was observed in the successive DBP ponds, which is in contrast with the results of the daytime period. Meanwhile, there was no consistent trend in the AFP ponds.

CH₄ fluxes
Both DBP and AFP behaved as sources of CH₄ during daytime and night-time (**Figure 4.4**). For daytime conditions CH₄ median emissions were 9.9 mg m^{-2} d^{-1} and 71.4 mg m^{-2} d^{-1} for the AFP and DBP, respectively. During night-time conditions AFP median emissions were 12.7 mg CH₄ m^{-2} d^{-1}, and DBP median emissions were 195.2 mg CH₄ m^{-2} d^{-1}. Thus, CH₄ fluxes in the DBP were higher under night-time conditions than under daytime conditions ($p = 0.027$), whereas there were no differences for the AFP CH₄ fluxes ($p>0.05$). In addition, median CH₄ emissions from the AFP and DBP ponds were 72.1 and 178.9 mg.CH₄.m^{-2}.d^{-1}, on a 24-hour basis, respectively.

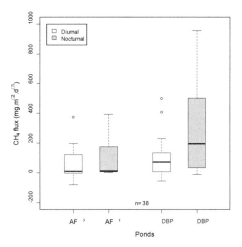

Figure 4.4 Fluxes of CH₄ from the AFP and DBP measured during daytime and night-time periods.

As shown in **Figure 4.5a** and **4.5b**, the CH₄ emissions decreased in the successive cylindrical reactors of the DBP and AFP ponds, during daytime and night-time periods. The results showed the highest CH₄ emissions occurred in the first tank (DBP1 and AFP1), and the lowest in the third tank (DBP3 and AFP3). This suggests a spatial variation in these emissions within the ponds.

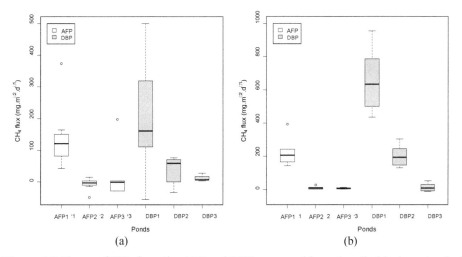

(a)   (b)

Figure 4.5 Fluxes of CH₄ from the AFP and DBP measured in each cylindrical reactor during daytime (a) and night-time (b) periods.

A comparison of CH_4 fluxes amongst the DBP and AFP (for instance, DBP1 and AFP1 and so on) under daytime conditions showed that there were no significant differences between them (p>0.05). On the other hand, the same comparison under night-time conditions showed that significant (p<0.05) differences occurred between the CH_4 fluxes from the first and second AFP and DBP tanks, whereas the CH_4 emissions from the third tank were not significantly different for either the AFP or the DBP, respectively

N_2O fluxes
The entire median N_2O fluxes were positive (**Figure 4.6**) suggesting that both ponds could be considered as an emission source of this gas. The values of N_2O fluxes during the day ranged from -4.1 to 63.5 mgm^{-2} d^{-1} for the AFP (Median = 6.9; SD = 16.5; n = 18) and between 1.6 and 42.3 mg N_2O m^{-2} d^{-1} for the DBP (Median = 8.5; SD = 10.7; n = 19). For the AFP, the N_2O fluxes during the night-time were in the range between -8.8 and 40.5 mg m^{-2} d^{-1} (Median = 5.5; SD = 12.8; n = 16), and the DBP ranged from -16 to 23.8 $mg.m^{-2}.day^{-1}$ (Median = 2; SD = 10.7; n = 18) (see again **Figure 4.6**).

Statistically, the differences between day and night of the AFP N_2O fluxes were not significant (p = 0.128), while fluxes from the DBP showed significant differences (p = 0.006). However, when the AFP daylight was compared with the DBP daylight and nocturnal AFP fluxes with diurnal DBP fluxes, there were no significant differences, which means that both ponds showed similar behaviour with respect to N_2O fluxes.

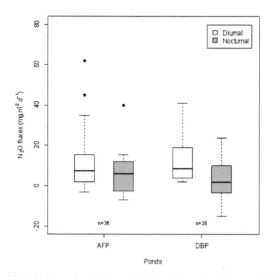

Figure 4.6 Fluxes of N_2O from the AFP and DBP during daytime and night-time

The behaviour of N_2O median fluxes in the successive tanks (**Figure 4.7a**) suggested similar pattern for AFP1, AFP2 and AFP3, and DBP1, DBP2 and DBP3 under daytime, although median values were slightly higher for AFP2 and DBP2. The comparison of AFP1 in relation to AFP2, and AFP3 fluxes showed that there were no significant differences between them (p = 0.368), whereas DBP1 was significantly different from DBP2 and DBP3 fluxes (p = 0.001). N_2O consumption was observed at AFP1 and AFP2 but it was only 2% of the total fluxes. In DBP1, DBP2 and DBP3 there was no N_2O consumption.

The median N_2O fluxes (**Figure 4.7b**) were significantly higher from AFP2 during the night-time compared to AFP1 and AFP3 (p<0.05). No statistically significant difference in flux rates was detected between DBP1, DBP2, and DBP3 (p>0.05). There were no significant differences between AFP1 and DBP1, thus these ponds were operating similarly in relation to N_2O fluxes. For AFP3 negative fluxes indicating N_2O consumption, were recorded during night-time operation.

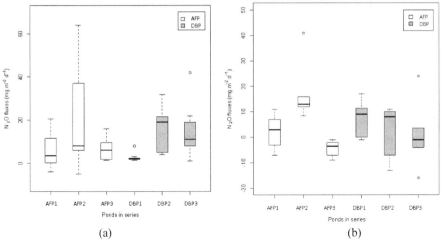

(a)   (b)

Figure 4.7 Fluxes of N_2O from the AFP and DBP measured in each cylindrical pond during the daytime (a) and night-time (b). In the whiskers are showed the quantile: Q1, Q2, and Q3; Q2 = Median; IRQ = Q3 - Q1; Min = Q1 - 1.5 IRQ; Max = Q3 + 1.5 IRQ.

4.4 DISCUSSION

4.4.1 CO$_2$ dynamics

In order to place the fluxes measured from the AFP and DBP and others studied in a broad perspective they were compared with CO_2 fluxes from FWS, HSSF and VSSF constructed wetlands and natural systems i.e. lakes and reservoir. The CO_2 flux rates obtained in this study were comparable to the average summer fluxes from FWS constructed wetland located in Norway (Sovik and Klove, 2007), and Finland (Søvik *et al.*, 2006) which were 1,200 and 1,400 mg.m^{-2}.day^{-1}, respectively. The cited studies quantified CO_2 fluxes from shallow (vegetation presence) and deep (open water) zones. Likewise, the 24-hour median emissions in our study were comparable to those measured in HSSF constructed wetlands treating young and stabilized leachate (Chiemchaisri et al., 2009), and HSSF mesocosms where it was observed that CO_2 fluxes were higher in planted and non-aerated units (Maltais-Landry et al., 2009). On the other hand, Liikanen et al. (2006) found 5 times higher CO_2 rate fluxes for HSSF and FWS constructed wetlands purifying peat mining runoff waters operating at higher loading rates of organic matter (490 mg.m^{-2}.d^{-1} COD) than in this study (13.9 mg.m^{-2}.d^{-1} COD). The median CO_2 flux were 1.5 times lower than measured on HSSF and VSSF constructed wetlands for wastewater treatment (Teiter and Mander, 2005). A much higher range (1,390-77,500 mg.m^{-2}.d^{-1}) of flux rates was reported by Ström et al. (2007). In the last two studies mentioned the CO_2 emissions were not connected with fluxes related to plant photosynthesis as in our study. The ranges and median of CO_2 flux found in our study also showed a magnitude comparable to those observed in reservoirs and natural systems (Casper et al., 2000; St. Louis et al., 2000; Huttunen et al., 2003; Tremblay et al., 2004; Xing et al., 2005; Roland et al., 2010).

The CO_2 dynamic in lakes, reservoirs, natural and constructed wetlands have been explained by air and soil temperature, depth of the water table, and vegetation being the most important factors controlling CO_2 fluxes. A higher CO_2 rate production was observed in constructed wetlands (CWs) when air and sediment temperature were increased (Huttunen et al., 2002; Teiter and Mander, 2005; Liikanen et al., 2006; Søvik et al., 2006). In boreal lakes, the CO_2 release was higher from the low than from the high water table (Larmola et al., 2003; Liikanen et al., 2003). In contrast the level of the water table did not affect the CO_2 fluxes in a CWs purifying peat mining runoff waters (Liikanen et al., 2006). In constructed wetlands, the CO_2 fluxes were higher in the vegetated areas than in the non-vegetated areas (Søvik et al., 2006; Picek et al., 2007; Ström et al., 2007). This influence of vegetation on CO_2 emissions was related to the importance of vegetation in the gas transportation, primarily by pressurized convective gas flow (Brix et al., 2001). However, the differences between daytime and night-time CO_2 found in our study suggest that in addition to the factors previously mentioned, the CO_2 emissions from the AFP and the DBP are strongly influenced by photosynthetic activity of algae and duckweed, heterotrophic respiration, and dark respiration from plants.

The net carbon dioxide release from the AFP and DBP is the difference between the respiratory processes producing CO_2, and the photosynthetic processes fixing it. Thus, during daylight conditions the CO_2 fixation carried out by algae and plants would exceed the heterotrophic respiration and the systems worked as sinks (Xing et al., 2005). In addition, during daylight the dissolved CO_2 (aq.) is depleted due to photosynthesis causing the pH of water to increase. At pH exceeding 8, the formation of carbonic acid and bicarbonate leads to an under-saturation of dissolved CO_2 in the water column and enhances the net transfer of CO_2 from the atmosphere into water. In contrast, at night algae and plants switch to respiration and hence produce CO_2 in addition to the CO_2 produced by heterotrophic bacteria, which may lead to CO_2 super-saturation in the water column. Thus, during nocturnal periods the ponds behaved as net carbon dioxide sources.

The results found in this study suggest that uptake of CO_2 by photosynthesis should be considered to estimate the CO_2 fluxes from AFPs and DBPs. The uptake CO_2 rate by photosynthesis has been shown as an important fraction of carbon fluxes in aquatic ecosystems (Brix *et al.*, 2001; Teiter and Mander 2005; Ström *et al.*, 2007). Therefore, not taking into account the photosynthetic activity could overestimate the emissions from AFPs, DBPs and other aquatic ecosystems.

The CO_2 fluxes varied between the cylindrical tanks in both the AFP and DBP during daytime and night-time periods. We found that AFP1 and DBP1 had different operational characteristics than successive tanks (AFP2, AFP3 or DBP2, DBP3). Since the first tank received UASB effluent probably the conditions for algae and duckweed were affected. This was corroborated by chlorophyll-a median concentration values measured in AFP1 (30 µg l^{-1}), which were lower than AFP2 and AFP3, at 1,200 and 1,800 µg l^{-1}, respectively. In addition, the duckweeds looked unhealthy and had growth difficulties in the DBP1 tank and therefore its density was lower (<400 gm^{-2}) than observed in the other successive ponds (>700 gm^{-2}). This is likely due to higher process rates at higher organic matter concentrations in the inlet or extension of anaerobic conditions in this tank (AFP1 or DBP1). This pattern has also been observed in studies about HSSF constructed wetlands (Søvik *et al.*, 2006; Picek *et al.*, 2007) which report that carbon dioxide emissions from the inlet section were higher than the emissions from the outlet section.

As AFP1 and DBP1 were different than the other tanks, only AFP2, AFP3 and DBP2, DBP3 were the tanks that represent real algae and duckweed pond behaviour. To determine the daytime and night-time CO_2 fluxes from real algae and duckweed ponds, the median of the fluxes from AFP2- AFP3 together and DBP2-DBP3 together were calculated. According this, the daylight median CO_2 fluxes measured in AFP2-AFP3, and DBP2-DBP3, were -1,034 mg CO_2.m^{-2}.d^{-1} and -863 mg. CO_2.m^{-2}.d^{-1} respectively. On the other hand, during nocturnal periods the median CO_2 fluxes for AFP2-AFP3 values and DBP2-DBP3 were 4,570.9 mg CO_2.m^{-2}.d^{-1} and 4,322 mg CO_2.m^{-2}.d^{-1} respectively. These figures indicated that under daytime conditions AFP2-AFP3 showed stronger sink behaviour than DBP2-DBP3 while during nocturnal periods AFP2-AFP3 emitted more CO_2 than DBP2-DBP3.

4.4.2 CH$_4$ dynamics

None of the previous studies found in the literature were related directly to CH$_4$ fluxes from algae or duckweed ponds. However, our results were similar to values reported for HSSF and VSSF constructed wetlands operating in Europe where mean fluxes were 110 mgCH$_4$.m^{-2}.d^{-1} (Søvik *et al.*, 2006) and 141 mgCH$_4$.m^{-2}.d^{-1} (Johansson *et al.*, 2004). The CH$_4$ mean fluxes were lower than most of the values reported in the literature for HSSF and FWS constructed wetlands, e.g. 225 $mgCH_4.m^{-2}.d^{-1}$ (Tanner et al., 1997), 240 $mgCH_4.m^{-2}.d^{-1}$ (Stadmark and Leonardson 2005), 400 $mgCH_4.m^{-2}.d^{-1}$ (Liikanen et al., 2006) and 343 $mgCH_4.m^{-2}.d^{-1}$ (Chiemchaisri et al., 2009), but higher than values reported by Teiter and Mander, (2005) which was 7.39 $mgCH_4.m^{-2}.d^{-1}$. In addition, the CH_4 rate fluxes found in this study were higher than those reported for natural wetlands and lakes (Casper et al., 2000; Huttunen et al. 2002; Saarnio et al., 2009), reservoirs (St. Louis et al., 2000; Guérin et al., 2006), and forested riparian wetlands (Hopfensperger et al., 2009). The latter is logically expected as AFP and DBP ponds for wastewater treatment are eutrophic ecosystems whilst natural well-preserved ecosystems are mostly oligotrophic.

Methane emission behaviour in the WSP studied can be explained by three processes: *i*) separation of remaining dissolved methane from the UASB pre-treatment (Cakir and Stenstrom, 2005), *ii*) anaerobic processes of the remaining biodegradable (particulate or bio-flocs) organic fraction in the wastewater (El-Fadel and Massoud, 2001) and *iii*) methane oxidation and methanogenesis inhibition (Huttunen et al., 2002; Johansson et al., 2004).

Regarding the first process, the dissolved methane from the UASB effluent may have caused the CH_4 emissions to be higher in the first tank i.e. AFP1 and DBP1 (HRT = 4 days) of both ponds during daytime and night-time (**Figure 4.5a** and **4.5b**). The methane micro-bubbles are so tiny they are suspended in the liquid and then enter into the ponds where these would separate due to the relatively long HRTs and diffusive effects. In addition, CH_4 is considered insoluble in water and the absence of physical barriers (i.e. duckweed coverage) or high algae concentrations that supply oxygen for methane oxidation favoured the CH_4 emissions from AFP1 and DBP1. In this case, the methane release from AFP1 and DBP1 is transported by ebullition and diffusion into the atmosphere (Johansson et al., 2004; Sovik and Klove, 2007).

According to the second process, the anaerobic biodegradable matter decomposition may also add to explaining the high CH_4 emissions measured in the AFP1 and DBP1 ponds. At the inlet of these first ponds higher organic loading was compared with subsequent ponds. According to this, AFP1 and DFP1 did not really function as full algal and duckweed ponds; they were instead functioning as over-loaded secondary facultative or duckweed ponds. The higher methane emissions from the inlet section in CW have been attributed to higher process rates at higher carbon concentrations in the inlet (Søvik et al., 2006).

The third process mentioned earlier may better explain the methane emissions from the successive pond units in the series (AFP2, AFP3, DBP2 and DBP3). In the current study were found super saturation concentrations of DO (mean of 13.2 $mgO_2.L^{-1}$) during daytime conditions in AFP2 and AFP3. This is related to intense algal activity as shown by the high chlorophyll-a concentration (median 1,296 $\mu g.l^{-1}$). This indicates a high methane oxidation potential in these reactors that contributed to negative fluxes of this gas (median -2.8 $mgCH_4.m^{-2}.d^{-1}$). Ying et al. (2000) attributed the decrease in CH_4 emissions to an increase in DO in a rice wetland and Huttunen *et al.* (2002) and Liikanen & Martikainen (2003) stated that well-oxygenated water columns in shallow ponds can limit methanogenesis in the sediments and favours methane oxidation instead. According to Johansson *et al.* (2004), methane emissions from constructed wetlands could be efficiently reduced through management practices that favour the establishment of algae such as *Spirogyra* sp.

In contrast, during the night-time, AFP2 and AFP3 showed CH_4 positive fluxes (median 7.1 $mgCH_4.m^{-2}.d^{-1}$) and therefore were net sources of this gas. In nocturnal conditions, algae switch to respiration, which leads, in combination with the activity of heterotrophic bacteria, to a faster reduction of the remaining dissolve O_2. Meanwhile, DO concentrations in AFP2 and AFP3 during night-time conditions reached median values of 1.8 mg. O_2 l^{-1}, which probably reduces the methane oxidation potential. Reducing conditions were also observed for AFP2 and AFP3 (median ORP = -8mV), which is an indication of weak anoxic conditions on set, thus favouring methane formation.

The availability of oxygen is not the only factor that causes methane oxidation, for instance CH_4 oxidation is dependent upon sulphates (Valentine, 2002), temperature, concentration and nitrogen source (Nozhevnikova *et al.*, 2001). Stadmark and Leonardson (2007) hypothesized that at higher NO_3^--N concentrations, CH_4 production is inhibited. Thus, as we observed nitrate concentrations similar to those reported by these authors in AFPs —regularly higher compared to the DBPs — this fact could explain the lower CH_4 emissions from the AFP observed in nocturnal periods. However, when a statistical model was applied to determine the influence of NO_3^- - N on CH_4 emissions, no correlation of that sort was found.

On the other hand, the DBP2 and DBP3 CH_4 fluxes were positive in daytime and night-time conditions; thus, these tanks can be considered as moderate sources of CH_4. The low DO concentrations (median DO 1.44 $mg.O_2 L^{-1}$) and negative ORP conditions (median ORP = -260) observed in DBP2 indicated anoxic conditions in this tank. This probably favoured more conversion to methane in the anoxic sediment layer. In DBP3 a considerable proportion of the CH_4 fluxes were negative (60% of all the fluxes) suggesting that DBP3 behaved as a sink. This could be explained by considering that DBP3 had less organic matter and therefore less methane is produced by methanogenesis than in the previous ponds. Additionally, the oxygen concentrations were higher and therefore the methane oxidation potential also increased. The higher biomass density of floating duckweed in DBP3 may have functioned as a physical barrier for gas transfer, thus preventing a major release of CH_4 as suggested previously by Van der Steen *et al.* (2003b).

4.4.3 N_2O dynamics
The N_2O flux rates showed a similar behaviour for the AFP and DBP, leading to the average of N_2O fluxes during 24 hours for the AFP and DBP of 9.5 and 8.6 $mg.N_2O.m^{-2}.d^{-1}$, respectively. These mean values were higher than those found in constructed wetlands, e.g. 3.20 $mg.N_2O.m^-$

$^2.d^{-1}$ (Fey *et al.*, 1999), 3.12 mg.N_2O.m^{-2}.d^{-1} (Johansson *et al.*, 2003), 0.82 mg.N_2O.m^{-2}.d^{-1} (Teiter and Mander, 2005), 0.45 mg.N_2O.m^{-2}.d^{-1} (Liikanen *et al.*, 2006), 5.9 mg.N_2O.m^{-2}.d^{-1} (Søvik *et al.*, 2006), and 5.5 mg.N_2O.m^{-2}.d^{-1} (Ström *et al.*, 2007). Sovik and Klove (2007) found higher N_2O fluxes in a constructed wetland in Norway. Sing *et al.* (2005) found in a tropical urban pond fluxes varying between 0.00 and 0.51 mg.N_2O.m^{-2}.d^{-1}. Picek *et al.* (2007) reported that in a HSSF treating municipal wastewater the N_2O emission was negligible and the only product of denitrification was N_2. Likewise, the median N_2O fluxes reported for natural systems were lower than those obtained in this study (Huttunen *et al.*, 2002; Huttunen *et al.*, 2003; Hopfensperger *et al.*, 2009).

N_2O fluxes from the successive AFP and DBP tanks can largely be attributed to differences in nitrification and denitrification activity. The conditions in AFP1 and DBP1, i.e. partial anaerobic conditions, availability of biodegradable organic substrate and moderate nitrate concentration, could be favourable for denitrification. Denitrification involves nitrate or nitrite reduction to molecular nitrogen, without accumulation of N_2O, which should only be considered as an intermediate stage of the respiratory process. Since N_2O was emitted from these tanks, it suggests that probably the denitrification process was incomplete.

A number of factors have been suggested to explain the cause of N_2O in denitrification process. The availability of nitrate probably promotes high N_2O emission by denitrification in constructed wetlands for wastewater treatment (Søvik *et al.*, 2006). Von Schulthess *et al.*, (1994) found that N_2O emission increased from denitrification due to increased oxygen concentrations. Hanaki *et al.* (1992) reported N_2O production during denitrification in the presence of low COD/N ratio (3.5), decreasing pH from 8.5 to 6.5 and short solid retention time (<1 day). According to Itokawa *et al.* (2001), more than 20% of influent nitrogen can be emitted as N_2O during biological nitrogen removal of high strength wastewater if COD/N ratio is below 3.5. Because the AFP1 and DBP1 tanks in our study were operated at a 3.2 COD/N ratio, pH of 7-8.4 and a DO average 1.5 mg.l^{-1} the N_2O, emissions probably are low due to the COD/N applied and nitrate presence in these tanks.

According to Zimmo *et al.* (2004b), in DBP ponds the nitrification is possible on the duckweed mat and denitrification mainly occurs in the sediment. Therefore, in DBP2 and DBP3, ammonium could be oxidized through nitrification to nitrate on the mat surface, which can then diffuse into the adjacent anoxic zones to be reduced in the sediment and thus to allow N_2O production. This suggests that the nitrification in a DBP is the prerequisite for N_2O emissions because denitrification is coupled to nitrification. The conditions of low temperature and low The DO observed in DBP2 and DBP3 during the night-time may have been responsible for lower nitrification and consequently less N_2O was emitted. In contrast, during daytime higher temperature should result in an increased nitrification and N_2O was largely emitted. Another factor that could explain the high N_2O emissions during daytime could be the intensive harvesting of duckweed (each 3 days) that probably reduced the possibility to prevent N_2O emissions into the atmosphere (Van der Steen *et al.* 2003b).

4.5 CONCLUSIONS

- This study has provided one of the very first sets of comprehensive CO_2, CH_4, and N_2O emission results for two pilot-scale shallow stabilization ponds (AFP and DBP) under tropical conditions. According to the results, the AFP and DBP under the studied conditions were net sources of CO_2, CH_4, and N_2O when taking into account a 24-hr period.

- The differences found between daytime and night-time CO_2, CH4 and N_2O fluxes suggest that the GHG dynamic in AFP and DBP is influenced by the photoperiod. This indicates that when estimating the CO_2, CH_4 and N_2O fluxes from either the AFP or DBP the fluxes should to be measured considering daytime and night-time measurements. Overall, when photosynthesis is omitted the fluxes from the AFP and DBP are probably overestimated and the consequence of this is a large uncertainty in GHG estimation from the AFP and DBP.

- In the ponds where the duckweed cover was well developed the CH_4 emissions were relatively low and, in some occasions, were negative (in DBP3). This suggests that the duckweed mat cover acts as a physical barrier to prevent the CH_4 emissions from the DBP.

- This study provides evidence that facultative ponds i.e. AFP and DBP impact the environment from point view GHG emissions.

4.6 ACKNOWLEDGMENTS

The authors acknowledge the financial assistance of the SWITCH project for the PhD fellowship of J.P. Silva, and COLCIENCIAS for the fellowship of J. Ruiz under the young researcher programme. The authors are also indebted to ACUAVALLE ESP and the Universidad del Valle for their support. We also recognize the contribution by the undergraduates in sanitary engineering, Angela Reyes and María Helena Dominguez, for their assistance.

4.7. REFERENCES

APHA, AWWA and WEF (2005) Standard methods for the examination of water and wastewater, American Public Health Association, American Water Works Association and Water Environment Federation, Washington DC.

Arthur, J.P. (1983) Notes on the design and operation of waste stabilization ponds in warm climates of developing countries. Technical Paper 7, The World Bank, Washington, DC.

Awuah, E., Oppong-Peprah, M., Lubberding, H.J. and Gijzen, H.J. (2004). Comparative Performance Studies of Water Lettuce, Duckweed, and Algal-Based Stabilization Ponds Using Low-Strength Sewage. . J. Toxicol. Environ. Health-Part A-Current 67, 1727.

Bogner, J., M. Abdelrafie Ahmed, C. Diaz, A. Faaij, Q. Gao, S. Hashimoto, K. Mareckova, R. Pipatti and Zhang, T. (2007). Waste Management, In Climate Change 2007: Mitigation. Contribution of Working Group III to the Fourth Assessment Report of the Intergovernmental Panel on Climate Change. B. Metz, O.R. Davidson, P.R. Bosch, R. Dave and Meyer, L.A. (eds), p. 32, Cambridge University, Cambridge, United Kingdom

Brix, H., Sorrell, B.K. and Lorenzen, B. (2001). Are Phragmites-dominated wetlands a net source or net sink of greenhouse gases? Aquatic Botany 69(2-4), 313-324.

Caicedo, J.R. (2005). Effect of Operational Variables on Nitrogen Transformations in Duckweed Stabilization Ponds. PhD, UNESCO-IHE- Wageningen University, Delft.

Cakir, F.Y. and Stenstrom, M.K. (2005). Greenhouse gas production: A comparison between aerobic and anaerobic wastewater treatment technology. Water research 39, 4197-4203.

Casper, P., Maberly, S.C., Hall, G.H. and Finlay, B.J. (2000). Fluxes of methane and carbon dioxide from a small productive lake to the atmosphere. Biogeochemistry 49(1), 1-19.

Chiemchaisri, C., Chiemchaisri, W., Junsod, J., Threedeach, S. and Wicranarachchi, P.N. (2009). Leachate treatment and greenhouse gas emission in subsurface horizontal flow constructed wetland. Bioresource technology 100(16), 3808-3814.

El-Fadel, M. and Massoud, M. (2001). Methane emissions from wastewater management. Environmental Pollution 114(2), 177-185.

El-Fadel, M. and Massoud, M. (2002). Methane from Wastewater Management. Environmental Pollution 114, 177-185.

El-Shafai, S.A., El-Gohary, F.A., Nasr, F.A., Peter van der Steen, N. and Gijzen, H.J. (2007). Nutrient recovery from domestic wastewater using a UASB-duckweed ponds system. Bioresource technology 98(4), 798-807.

Fey, A., Benckiser, G. and Ottow, J.C.G. (1999). Emissions of nitrous oxide from a constructed wetland using a groundfilter and macrophytes in waste-water purification of a dairy farm. Biology and fertility of soils 29(4), 354-359.

Forster, P., Ramaswamy, V., Artaxo, P., Berntsen, T., Betts, R., Fahey, D.W., Haywood, J., Lean, J., Lowe, D.C., Myhre, G., Nganga, J., Prinn, R., Raga, G., Schulz, M. and Van Dorland, R. (2007). Changes in atmospheric constituents and in radiative forcing. . In Climate Change 2007: the physical science basis. Contribution of Working Group I to the Fourth Assessment Report of the Intergovernmental Panel on Climate Change. Solomon, S., Qin, D., Manning, M., Chen, Z., Marquis, M., Averyt, K. B., Tignor, M. & Miller, H. L. (ed), Cambridge University Press, Cambridge, United Kingdom and New York, NY, USA.

Greenfield, P.F., Batstone, D.J., Guiot, S.R., Pavlostathis, S.G. and van Lier, J.B. (2005). Anaerobic digestion: impact of future greenhouse gases mitigation policies on methane generation and usage. Water Science and Technology 52(1-2), 39-47.

Guérin, F., Abril, G., Richard, S., Burban, B., Reynouard, C., Seyler, P. and Delmas, R. (2006). Methane and carbon dioxide emissions from tropical reservoirs: significance of downstream rivers. Geophysical Research Letters 33(21), L21407.

Hanaki, K., Hong, Z. and Matsuo, T. (1992). Production of nitrous oxide gas during denitrification of wastewater. Water Science and Technology 26(5/6), 1027-1036.

Hopfensperger, K.N., Gault, C.M. and Groffman, P.M. (2009). Influence of plant communities and soil properties on trace gas fluxes in riparian northern hardwood forests. Forest Ecology and Management 258(9), 2076-2082.

Huttunen, J.T., Alm, J., Liikanen, A., Juutinen, S., Larmola, T., Hammar, T., Silvola, J. and Martikainen, P.J. (2003). Fluxes of methane, carbon dioxide and nitrous oxide in boreal lakes and potential anthropogenic effects on the aquatic greenhouse gas emissions. Chemosphere 52(3), 609-621.

Huttunen, J.T., Väisänen, T.S., Heikkinen, M., Hellsten, S., Nykänen, H., Nenonen, O. and Martikainen, P.J. (2002). Exchange of CO_2, CH_4 and N_2O between the atmosphere and two northern boreal ponds with catchments dominated by peatlands or forests. Plant and Soil 242(1), 137-146.

IPCC (2006). 2006 IPCC Guidelines for National Greenhouse Gas Inventories. Eggleston H.S., B.L., Miwa K., Ngara T. and Tanabe K. (ed), ntergovernmental Panel on Climate Change, Hayama, Japan.

Itokawa, H., Hanaki, K. and Matsuo, T. (2001). Nitrous oxide production in high-loading biological nitrogen removal process under low COD/N ratio condition. Water research 35(3), 657-664.

Johansson, A.E., Gustavsson, A.M., Öquist, M.G. and Svensson, B.H. (2004). Methane emissions from a constructed wetland treating wastewater--seasonal and spatial distribution and dependence on edaphic factors. Water research 38(18), 3960-3970.

Johansson, A.E., Klemedtsson, Å., Klemedtsson, L. and Svensson, B.H. (2003). Nitrous oxide exchanges with the atmosphere of a constructed wetland treating wastewater. Tellus B 55(3), 737.

Johnson, M.L. and Mara, D.D. (2007). Ammonia removal from facultative pond effluents in a constructed wetland and an aerated rock filter: performance comparison in winter and summer. Water Environmental Research 79(5), 567-570.

Lambert, M. and Fréchette, J.L. (2005). Greenhouse Gas Emissions - Fluxes and Processes: Hydroelectric Reservoirs and Natural Environments Tremblay, A., Varfalvy, L., Roehm, C. and Garneau, M. (eds), pp. 37-60, Springer.

Larmola, T., Alm, J., Juutinen, S., Martikainen, P.J. and Silvola, J. (2003). Ecosystem CO2 exchange and plant biomass in the littoral zone of a boreal eutrophic lake. Freshwater Biology 48(8), 1295-1310.

Liikanen, A., Huttunen, J.T., Karjalainen, S.M., Heikkinen, K., Vaisanen, T.S., Nykanen, H. and Martikainen, P.J. (2006). Temporal and seasonal changes in greenhouse gas emissions from a constructed wetland purifying peat mining runoff waters. Ecological Engineering 26(3), 241-251.

Liikanen, A. and Martikainen, P.J. (2003). Effect of ammonium and oxygen on methane and nitrous oxide fluxes across sediment-water interface in a eutrophic lake. Chemosphere 52(8), 1287-1293.

Liikanen, A., Ratilainen, E., Saarnio, S., Alm, J., Martikainen, P.J. and Silvola, J. (2003). Greenhouse gas dynamics in boreal, littoral sediments under raised CO2 and nitrogen supply. Freshwater Biology 48(3), 500-511.

Maltais-Landry, G., Maranger, R. and Brisson, J. (2009). Effect of artificial aeration and macrophyte species on nitrogen cycling and gas flux in constructed wetlands. Ecological Engineering 35(2), 221-229.

Mara, D. (2005). Pond treatment technology. Shilton, A. (ed), pp. 168-187, IWA Publishing London.

Nalbur, B.E., Akca, L. and Bayhan, H. (2003). Nitrogen removal during secondary treatment by aquatic systems. Water science & technology 48(11), 355-361.

Nozhevnikova, A.N., Nekrasova, V.K., Kevbrina, M.V. and Kotsyurbenko, O.R. (2001). Production and oxidation of methane at low temperature by the microbial population of municipal sludge checks situated in north-east Europe. Water science and technology: a journal of the International Association on Water Pollution Research 44(4), 89.

Peña, M.R., Madera, C.A. and Mara, D.D. (2002). Feasibility of waste stabilization pond technology for small municipalities in Colombia. Water Science and Technology 45(1), 1-8.

Picek, T., Cizkova, H. and Dusek, J. (2007). Greenhouse gas emissions from a constructed wetland--Plants as important sources of carbon. Ecological Engineering 31(2), 98-106.

Roland, F., Vidal, L.O., Pacheco, F.S., Barros, N.O., Assireu, A., Ometto, J., Cimbleris, A.C.P. and Cole, J.J. (2010). Variability of carbon dioxide flux from tropical (Cerrado) hydroelectric reservoirs. Aquatic Sciences-Research Across Boundaries 72(3), 283-293.

Saarnio, S., Winiwarter, W. and Leitao, J. (2009). Methane release from wetlands and watercourses in Europe. Atmospheric Environment 43(7), 1421-1429.

Silva, J.P., Lasso, A., Lubberding, H.J., Peña, M.R. and Gijzen, H.J. (2015). Biases in greenhouse gases static chambers measurements in stabilization ponds: Comparison of flux estimation using linear and non-linear models. Atmospheric Environment 109, 130-138.

Singh, V.P., Dass, P., Kaur, K., Billore, S.K., Gupta, P.K. and Parashar, D.C. (2005). Nitrous oxide fluxes in a tropical shallow urban pond under influencing factors. Current Science 88(3), 478.

Søvik, A.K., Heikkinen, J., Huttunen, K., Necki, J.T., Karjalainen, J.M., Kløve, S.M., Liikanen, B., Mander, A., Puustinen, Ü. and Teiter, M. (2006). Emission of the greenhouse gases nitrous oxide and methane from constructed wetlands in Europe. Journal of environmental quality 35(6), 2360.

Sovik, A.K. and Klove, B. (2007). Emission of N2O and CH4 from a constructed wetland in southeastern Norway. Science of the Total Environment 380(1-3), 28-37.

St. Louis, V.L., Kelly, C.A., Duchemin, É., Rudd, J.W.M. and Rosenberg, D.M. (2000). Reservoir surfaces as sources of greenhouse gases to the atmosphere: A global estimate. BioScience 50(9), 766-775.

Stadmark, J. and Leonardson, L. (2005). Emissions of greenhouse gases from ponds constructed for nitrogen removal. Ecological Engineering 25(5), 542-551.

Stadmark, J. and Leonardson, L. (2007). Greenhouse gas production in a pond sediment: Effects of temperature, nitrate, acetate and season. Science of the Total Environment 387(1-3), 194-205.

Ström, L., Lamppa, A. and Christensen, T.R. (2007). Greenhouse gas emissions from a constructed wetland in southern Sweden. Wetlands Ecology and Management 15(1), 43-50.

Tanner, C.C., Adams, D.D. and Downes, M.T. (1997). Methane emissions from constructed wetlands treating agricultural wastewaters. Journal of environmental quality 26(4), 1056-1062.

Teiter, S. and Mander, U. (2005). Emission of N_2O, N_2, CH_4, and CO_2 from constructed wetlands for wastewater treatment and from riparian buffer zones. Ecological Engineering 25(5), 528-541.

Toprak, H. (1995). Temperature and organic loading dependency of methane and carbon dioxide emission rates of a full-scale anaerobic waste stabilization pond. Water research 29(4), 1111-1119.

Tremblay, A., Lambert, M. and Gagnon, L. (2004). Do hydroelectric reservoirs emit greenhouse gases? Environmental Management 33, 509-517.

UNFCCC (2007). Uniting on Climate. United Nations Framework Convention on Climate Change, Bonn.

Valentine, D.L. (2002). Biogeochemistry and microbial ecology of methane oxidation in anoxic environments: a review. Antonie Van Leeuwenhoek 81(1), 271-282.

Van der Steen, N.P., Ferrer, A.V.M., Samarasinghe, K.G. and Gijzen, H.J. (2003a). Quantification and comparison of methane emissions from algae and duckweed based wastewater treatment ponds. Universidad del Valle; CINARA; International Water Association. Memorias del evento: Agua 2003. Cartagena de Indias, IWA, 2003, p. 1-7 Ilus..

Van der Steen, N.P., Nakiboneka, P., Mangalika, L., Ferrer, A.V. and Gijzen, H.J. (2003b). Effect of duckweed cover on greenhouse gas emissions and odour release from waste stabilisation ponds. Water Sci Technol 48(2), 341-348.

Von Schulthess, R., Wild, D. and Gujer, W. (1994). Nitric and nitrous oxides from denitrifying activated sludge at low oxygen concentration. Water Science and Technology 30(6), 123-132.

Xing, Y., Xie, P., Yang, H., Ni, L., Wang, Y. and Rong, K. (2005). Methane and carbon dioxide fluxes from a shallow hypereutrophic subtropical lake in China. Atmospheric Environment 39(30), 5532-5540.

Ying, Z., Boeckx, P., Chen, G.X. and Van Cleemput, O. (2000). Influence of Azolla on CH4 emission from rice fields. Nutrient Cycling in Agroecosystems 58, 321-326.

Zimmo, O.R., Van der Steen, N.P. and Gijzen, H.J. (2004a). Nitrogen mass balance across pilot-scale algae and duckweed-based wastewater stabilisation ponds. Water research 38(4), 913-920.

Zimmo, O.R., Van Der Steen, N.P. and Gijzen, H.J. (2004b). Quantification of nitrification and denitrification rates in algae and duckweed based wastewater treatment systems. Environmental technology 25(3), 273-282.

Chapter 5

GHG emissions from algal facultative ponds under tropical conditions

This Chapter has been presented and published as:

Silva J.P., Caicedo F, Lubberding HJ, Peña MR, and Gijzen HJ (2011). GHG emissions from algal facultative ponds under tropical conditions SEP in: 11th IWA stabilization ponds, 21st -23rd March 2016, Leeds, UK.

Abstract

Full scale experiments were designed to enhance understanding of daytime and night-time variations of CH_4, CO_2 and N_2O levels in an algal facultative pond (AFP) under tropical conditions. The results showed that the AFP was a net source of CH_4 during both daytime ($2,466.8 \pm 989.8$ mg CH_4 m^{-2} d^{-1}) and night-time ($2,254 \pm 1,152.5$ mg CH_4 m^{-2} d^{-1}). The variations in CH_4 emissions were influenced by environmental factors and physicochemical parameters such as ambient temperature and total nitrogen ($r^2 = 0.52$; $p < 0.05$). For CO_2 emissions, a strong influence of the photoperiod was observed. During the daytime, the AFP served as a CO_2 sink (-743 ± 847.5 mg CO_2 m^{-2} d^{-1}) while at night-time it served as a CO_2 source ($2,497 \pm 1,334.8$ mg CO_2 m^{-2} d^{-1}). CO_2 production in the AFP was correlated positively to COD, and negatively to pH and DO. The significant difference between daytime and night-time CO_2 reflected changes in algal photosynthesis and heterotrophic respiration. N_2O fluxes from the AFP during daytime (-0.95 ± 2.7 mg N_2O m^{-2} d^{-1}) and night-time (3.8 ± 7 mg N_2O m^{-2} d^{-1}) showed significant differences. By regression analysis, N_2O fluxes showed a positive correlation with NO_3^- N, and TKN, and negative correlation to DO and COD.

Keywords: Greenhouse Gases, Ecotechnologies, Wastewater Stabilization Ponds, Algal Facultative Ponds

5.1 INTRODUCTION

Because of low operational costs and effective removal of polluting organic matter and pathogens, wastewater stabilization ponds (WSPs) are widely used as EWWT in developing countries (Arthur, 1983; Peña et al., 2002; Mara, 2005). However, the sustainable operation of WSPs could be affected by the high rate of conversion of organic and nitrogenous loads into greenhouse gases (GHG), mainly CH_4, CO_2 and N_2O. CH_4 and N_2O are especially very harmful to the environment. Because of their high global warming potential (GWP), they contribute enormously to the human-induced greenhouse effect (Forster et al., 2007).

The emissions of CH_4, CO_2 and N_2O have been measured in various types of WSPs (Toprak, 1995; Van der Steen et al., 2003; Singh et al., 2005; Stadmark and Leonardson, 2005). In these studies, the mechanisms for CH_4, CO_2, and N_2O emissions have been related to processes that involve methanogenesis, nitrification and denitrification. In addition, the sediment, water, and air temperature are environmental factors contributing to GHG dynamics in WSPs (Toprak, 1995; Singh et al., 2005). Likewise, the availability of substrate i.e. COD or nitrate limit CH_4 and N_2O production (Toprak, 1995; Johansson et al., 2003; Johansson et al., 2004; Stadmark and Leonardson, 2005; Søvik and Klove, 2007).

In AFPs, the CH_4, CO_2 and N_2O production under sub-tropical and seasonal conditions has been more frequently reported (Johansson et al., 2004; Stadmark and Leonardson, 2005;

Detweiler *et al.*, 2014; Hernandez-Paniagua *et al.*, 2014; Glaz *et al.*, 2016). In these studies, water temperature, turbulence, wind, photosynthetic activity by algae, microbiological processes and pond depth reportedly influenced the gas diffusion and ebullition through air-water interphase. Water temperature positively influences CO_2, CH_4, and N_2O fluxes from AFPs due to enhanced microbial activity (Singh *et al.*, 2005; Stadmark and Leonardson, 2005; Detweiler *et al.*, 2014). Under temperate conditions, however, highest emissions were observed during warmer summer months (Johansson *et al.*, 2004; Glaz *et al.*, 2016). Also limiting CH_4 and N_2O production is substrate availability i.e. COD and NO_3^-N (Johansson *et al.*, 2003; Johansson *et al.*, 2004; Stadmark and Leonardson, 2005). Inhibiting methane production are nitrate concentrations between 8 and 16 mg NO_3-NL^{-1} (Stadmark and Leonardson, 2005). A high uptake of N_2O in AFPs is known to occur when nitrate concentrations were low (<5 mg L^{-1}) (Johansson *et al.*, 2004) .

From the perspective of wastewater treatment, the role of AFPs in developing tropical countries and the potential adverse environmental impact on GHG emissions cannot be dismissed. In consideration of this concern, the objectives of the current study are (1) to determine the CH_4, CO_2 and N_2O fluxes from AFPs ponds under tropical conditions; (2) to compare daytime and night-time fluxes (light/dark influence); and (3) to determine the influence of environmental parameters such as pH, DO, and temperature on GHG fluxes from AFPs.

5.2 MATERIALS AND METHODS

5.2.1 Field site and measuring periods
Measurements of GHG were taken at a full-scale AFP located at an experimental research station for wastewater treatment in Ginebra, Colombia (3°43'25.98 N, 76°15'59.45 W). Fed exclusively by domestic wastewater, this AFP receives effluent from a full-scale anaerobic pond. In addition, the AFP operation occurred under steady state according to the conditions shown in **Table 5.1**.

Table 5.1 Design and operational characteristics of the AFP

Parameter	Units	AFP
Flow	$m^3.d^{-1}$	24
Depth	m	1.39
Volume	m^3	99
HRT	day	4.1
Organic loading	g $BOD_5.m^{-2}d^{-1}$	43.7

5.2.2 Wastewater sampling

The influent and effluent wastewater quality of the AFP was determined over 24-hour sampling campaigns subdivide into three 8-hour periods. During each period taking every hour a fixed-volume of sample and pooling all together for analysis created one composite sample. Additional wastewater grab samples were retrieved from the central point of the AFP, at the spot where GHG fluxes were measured, to assess the correlation between GHG fluctuations and wastewater characteristics.

Biochemical oxygen demand (BOD_5), chemical oxygen demand (COD), total suspended solids (TSS), alkalinity, total Kjeldahl nitrogen (TKN), ammonium nitrogen ($N-NH_4^+$), and nitrate nitrogen ($N-NO_3^-$) were measured according to Standard Methods (APHA *et al.*, 2005). Conductivity, pH, dissolved oxygen (DO), temperature and the oxidation-reduction potential (redox) were measured with electrodes. Chlorophyll-a was measured using an ultraviolet light *Aquafluor* (Turner Designs) handheld fluorometer.

5.2.3 Greenhouse gas fluxes measurements using closed static chambers

Weekly sampling campaigns were undertaken under maximum and minimum solar radiation conditions to measure CH_4, CO_2 and N_2O emissions. In Ginebra, 12:30 to 15:30 is the period of maximum solar radiation whereas 00:00 to 03:00 is the period of minimum radiation. Along with regular evaluations, sampling over a 7-day period was carried out to investigate the daily variability of GHG such as CH_4, CO_2 and N_2O. In keeping with this protocol, wastewater and GHG samples were taken at specific times: 00:00, 03:00, 06:00, 09:00, 12:00, 15:00, 18:00, and 21:00.

Greenhouse gas emissions (CH_4, CO_2, and N_2O) from the central point of the AFP were determined using closed static chambers technique (Silva *et al.*, 2015). The chamber was designed to minimize artefacts (tubes, rubber stoppers), which increase the possibility of gas leakage. The cylindrical chamber (0.48 m x 0.3 m: diameter x height) was constructed by cutting the top of a white polyethylene plastic container (**Figure 5.1**). The chamber was fitted with a serum stopper sampling port and a PVC tube connected to a pressure-release valve. The valve was kept open when the chamber was placed on the water and closed before and during measurements.

Using a special syringe, CH_4, CO_2 and N_2O sampling were taken over a 45-minute period at 15-minute intervals (0, 15, 30, 45 min) from a sampling port atop each chamber. The relatively insignificant 15-mL sample volume did not affect the concentration build up in the static chamber. An internal standard, to compensate for gas losses in the entire procedure, was therefore unnecessary. Finally, the gas samples were withdrawn directly through a needle into pre-evacuated containers of 5 mL volume.

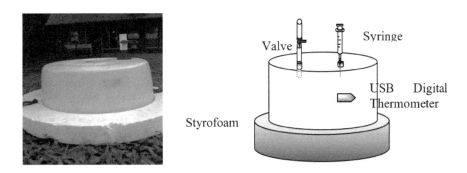

Figure 5.1 Static closed chamber used to measure GHG fluxes from an AFP.

CO_2. CO_2 was measured by an infrared spectrophotometer Qubit S151 CO_2 analyser (Loligo Systems, Denmark) using 75 mL min^{-1} air as the mobile phase with a temperature of the injector set equal to the ambient temperature.

CH_4. CH_4 was analysed by gas chromatography (Shimadzu Co., Japan) equipped with a flame ionic detector (FID). The Porapak Q column (80-100 mesh), was 2 m long and 2 mm in internal diameter. The temperatures at the injector, column, and detector were 80, 70, and 320° C, respectively. Carrier gas (N_2) flow rate was 22 mL min^{-1}.

N_2O. N_2O concentration was analysed similarly except that the equipment was fitted with an electron capture detector (ECD).

All samples were measured within 2 weeks after sampling. After every 10 samples the system was calibrated. In all cases Scotty standard gases were used for calibration (500 ppmv CO_2, 10 ppmv CH_4, 1 ppmv N_2O).

5.2.4 Flux estimations

In this paper, the term flux is defined as gas transfer per surface area ($ML^{-2}T^{-1}$) and the term emission is used for positive fluxes, while the term consumption is used for negative fluxes. The fluxes were calculated using equation 6.1 from linear and non-linear changes in the gas concentration in the closed static chamber headspace (Silva *et al.*, 2015).

$$F = \frac{dC}{dt_{t=0}} \times \frac{V_c}{A} \times \frac{1440 \ min}{d} \quad (Eq. \ 5.1)$$

F= the flux of CH_4, CO_2, N_2O ($gm^{-2}d^{-1}$); dC/ $dt_{t=0}$ = slope of the gas concentration curve ($gm^{-3}min^{-1}$); V_c= volume of the chamber (m^3); A=the cross-sectional area of the chamber (m^2).

5.2.5 Statistical Analysis

Statistical analyses were done with SPSS® software (v. 17.0 for Windows). The Kolmogorov-Smirnov test was used to check the normality of the data. An ANOVA analysis and the Wilcoxson test ($\alpha = 0.05$) were used to assess the daytime and night-time gas fluctuations. In addition, regression models were applied to determine the relation between wastewater characteristics and GHG variations. The models were verified for normality, independence, and homoscedasticity of residuals. Normality was checked by probability graphs P-P, and the independence of the Durbin Watson test. Homoscedasticity was carried out using a graph for the dispersion of residuals. Finally, the predicted values were standardized.

5.3 RESULTS

5.3.1 Performance of the AFP

The organic content of the AFP influent wastewater was low-strength (Metcalf and Eddy, 2003). **Table 5.2** shows the parameters in the influent, effluent and removal quantified in the AFP.

Table 5.2 Wastewater characteristics (mean ± SD) for influent and effluent AFP

Parameter (mg.l⁻¹)	Influent	Effluent	Removal %
COD $_{soluble}$	123 ± 59 (n=25)	87.0 ± 40.4 (n=25)	29.3
TSS	85.4 ± 69.8 (n=45)	64.3 ± 52.1 (n=45)	24.7
N – NO$_3^-$	1.1 ± 1 (n=30)	0.75 ± 0.4 (n=30)	31.8
TKN	40.7 ± 5.8 (n=30)	29.3 ± 3.9 (n=30)	28.0
N – NH$_4^+$	30.2 ± 4.5 (n=47)	21.4 ± 4.3 (n=45)	29.1

Analysis of the data revealed significant divergences between daytime and night-time ambient temperature, ORP, and dissolved oxygen ($p<0.05$) (**Table 5.3**). Measurement of other environmental parameters did not disclose marked deviations for the two periods studied.

5.3.2 GHG flux dynamics

CO₂ and CH₄ fluxes

The **Figure 5.2** displays the daytime and night-time fluxes for CO_2 and CH_4 in the AFP studied. During the daytime, the AFP acted as a CO_2 sink with fluxes ranging from -3,100 to 300 mg CO_2 m^{-2} d^{-1} (Mean= -743; SD=847.5; n=19). At nights, the AFP behaved as an emission source with values ranging from 1,040 to 4,730 mg CO_2 m^{-2}.d^{-1} (Mean=2,497; SD=1,334.8; n=19). Statistically, there were substantial differences between the CO_2 daytime and night-time gas fluxes (p=0.001).

As for methane, the daytime variances ranged from 290 to 4,510 mg CH_4 $m^{-2}.d^{-1}$ (Mean=2,466.8; SD=989.8; n=19) whereas night-time fluxes oscillated from 80 to 3,850 mg CH_4 $m^{-2}.d^{-1}$ (Mean=2,254; SD=1,152.5; n=19). ANOVA and Wilcoxon analyses showed no critical differences between daytime and night-time CH_4 fluxes (p=0.235) suggesting that the AFP was a net source of methane emission over a 24-h cycle

Table 5.3 Environmental parameters at the sampling point during GHG measurements. The water samples (n=19) were taken at 10 cm depth in the water column.

Parameter	Diurnal		Nocturnal	
	Mean	SD	Mean	SD
Water temp (°C)	28.3	2.7	24.8	1.4
Ambient temp (°C)	29.1	4.5	21.4	1.5
ORP (mV)	290.5	79.2	76.4	195.5
pH (Units)	8.5	0.5	7.1	0.6
DO (mg.l^{-1})	16.9	4.0	1.2	0.9
COD (mg.l^{-1})	224.1	139.8	226.4	36.8
TKN (mg.l^{-1})	30.5	6.1	33.4	14.9
N-NH$_4^+$ (mg.l^{-1})	20.5	3.0	17.9	1.9
N-NO$_3^-$ (mg.l^{-1})	1.6	0.2	1.4	0.2

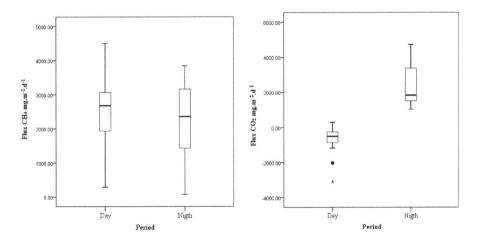

Figure 5.2 Daytime and night-time fluxes from the Ginebra AFP for CH_4 (a) and CO_2 (b)

The variation in CO_2 and CH_4 at the central point of the AFP in relation to the different environmental parameters was analysed using a regression model with data inputs from **Table 5.4**. The findings indicate that CO_2 fluxes correlated positively with COD and negatively with pH and DO. The model, therefore, clarifies the 74% variation in CO_2 distribution. An analysis of the individual contribution for each environmental parameter indicated that pH and DO fluctuations accounts for 37.5% and 31% respectively of the CO_2 variations. COD nonetheless could only account for 5.5% (**Table 5.4**).

Table 5.4 Linear regression analyses of CH_4 and CO_2 vs environmental parameters. TN, Ta, COD and DO denote total nitrogen, ambient temperature, chemical organic demand and dissolved oxygen.

Flux	Equation	r^2
CH_4	Flux CH_4 = 65.19 TN + 100.24Ta + 2322.6 + ε	0.52
CO_2	Flux CO_2 = 4.707 COD - 834.8 pH – 255.33 DO - 4130.78 + ε	0.74

CH_4 positively correlated to TKN and ambient temperature. However, these environmental parameters could only explain 52% of the CH_4 variations found in AFPs. The statistical analysis revealed that TKN accounted for 38.7% variations in methane gas, while ambient temperature accounted for 13.3%. Regrettably, no plausible hypothesis for the influence of temperature on CH_4 changes in the AFP can be adduced.

Daily CO_2 changes are depicted in **Figure 5.3** CO_2 fluxes were markedly lower during daytime than night-time reaching a minimum between 12:00 and 15:00 hours (-468 mg $m^{-2}.d^{-1}$). This coincided with the highest dissolved oxygen concentration (14.1 mg.l^{-1}), air temperature (34°C) and water alkalinity (pH 9.0). By contrast, the nightly (18:00 to 06:00) CO_2 released into the atmosphere peaked at of 3,400 mg $m^{-2}.d^{-1}$) and correlated with the minimal values of dissolved oxygen (1.2mg.l^{-1}) air temperature (20°C) and hydrogen ion concentration (pH 6.9).

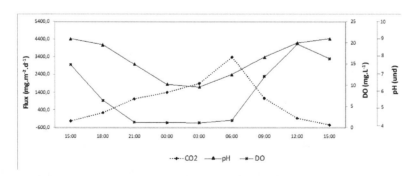

Figure 5.3 CO_2 fluxes from the Ginebra AFP during different time periods of the day in relation to DO and pH measured at the central point of the pond.

Figure 5.4 shows the daily cycles of CH₄ oscillations in the AFP. Although CH₄ measurements declined between 12:00 and 18:00 hours the CH₄ values remained positive implying that the AFP was a CH₄ source. At 18:00 hours, the CH₄ flux was around of 1,800 mg m^{-2}·d^{-1}. This level then soared to a peak of 3,800.6 mg m^{-2}·d^{-1} before levelling off. Coincidentally, the highest CH₄ emission was seen during with lowest level of dissolved oxygen and ORP condition (2mg.L^{-1} and 50 mV, respectively).

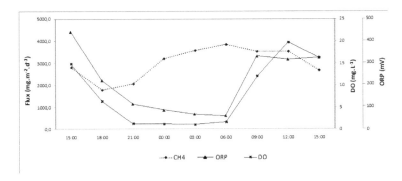

Figure 5.4 CH₄ fluxes from the Ginebra AFP during different time periods of the day in relation to DO and Oxidation-Reduction Potential (ORP) measured at the central point of the pond.

N₂O Fluxes
The N₂O diurnal fluxes (**Figure 5.5**) ranged from -6.3 to 3.6 mg m^{-2} d^{-1} (Mean= -0.95; SD=2.7; n=12) whereas the nocturnal fluxes ranged from -2.4 to 15 mg m^{-2} d^{-1} (mean 3.8; SD=7; n=12). In general, however, AFP studied was vast source of N₂O. Moreover, statistical analysis revealed a marginal but significant difference between day and night-time fluxes (p=0.049).

Regression was used to identify the correlation between environmental factors and N₂O fluxes. The nitrate-N (r^2=0.28), TKN (r^2=0.16), DO (r^2=0.14), COD (r^2=0.14) ambient temperature (r^2=0.12), water temperature (r^2=0.06), and ammonium-N (r^2=0.005) explained 85% of the variations in N₂O flux (**Table 5.5**).

Table 5.5 Models derived from a multiple stepwise regression of N₂O vs environmental parameters. N-NO₃, TN, Ta and Tw denote nitrate, total nitrogen, ambient temperature and water temperature.

Flux	Equation	r^2
N₂O	Flux N₂O = 954.8 N-NO₃ + 0.16 TN + 0.996 Ta – 0.97 Tw – 0.41 DO – 75.6 N-NH₄ – 0.05 COD + 16.796 + ε	0.85

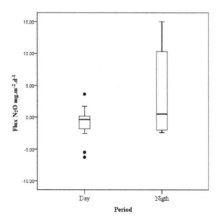

Figure 5.5 Comparison of N$_2$O fluxes from the AFP measured during day and night

N$_2$O showed daily fluctuations at different time periods (**Figure 5.6**). Two peaks of N$_2$O emissions were observed: at 15:00 hours N$_2$O emissions rose from 3.5 mg m^{-2} d^{-1} and peaked at 7.3 mg m^{-2} d^{-1}. Following a decline to -0.5 mg m^{-2} d^{-1} between 3:00 and 6:00 hours, emissions then climbed steadily reaching a second peak of 4.3 mg m^{-2} d^{-1} at 9:00 hours. Ultimately, they equilibrated at 0.8 mg m^{-2} d^{-1} between 9:00 and 15:00.

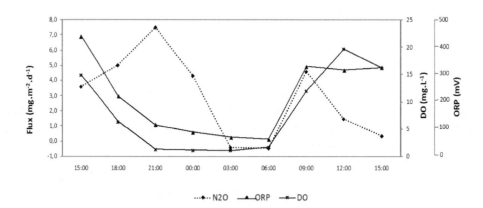

Figure 5.6 N$_2$O fluxes during different time periods of the day

The findings clearly reflect significant net N$_2$O emissions from the AFP despite the consumption observed between 3:00 to 6:00 hours (night-time). Regardless of the mechanisms involved, when the N$_2$O variation is compared to DO concentration and ORP profiles, sometime during the night when the oxygen tension is depleted and ORP has waned, N$_2$O emissions soar.

5.4 DISCUSSION

5.4.1 CO_2 dynamics

Testing for GHG showed that AFP studied was a night-time source and daytime sink for CO_2 (**Table 5.6**). Of significance is that the daytime CO_2 sink-effect does not substantially offset emissions produced by heterotrophic respiration resulting in the Ginebra's AFP being a net CO_2 producer. These findings are in line with those reported for secondary facultative ponds located in boreal and Mediterranean zones where high phytoplankton biomass also was responsible for regulating the overall pattern of consumption and production of CO_2 (Glaz *et al.*, 2016). By contrast, there is no evidence of similar consumption in subtropical SAFPs (Hernandez-Paniagua *et al.*, 2014) and duckweed ponds (Sims *et al.*, 2013; Dai *et al.*, 2015). Because of the conflicting evidence more research on the contribution of CO_2 by AFPs is needed.

The CO_2 produced by AFP may vary daily depending on their innate characteristics and environmental factors such as pH and temperature. The source- sink-effect of AFP on CO_2 is governed by the balance between photosynthetic algal biomass and heterotrophic respiration (Mara, 2005). Daytime algal photosynthesis in AFP assimilates CO_2 from pond water faster than atmospheric CO_2 diffusion and CO_2 released from bacterial degradation and the final outcome is a net consumption of CO_2 ((Teiter and Mander, *2005;* Picek *et al., 2007;* Shilton *et al.*, 2008). However, AFP can be a nocturnal source of CO_2 emissions due to algal heterotrophic respiration (Silva *et al.*, 2012).

In the Ginebra AFP a negative correlation was found between pH and CO_2 emissions. At pH values above 8.0 the formation of carbonic acid and bicarbonate leads to an under-saturation of dissolved CO_2 in the water column enhancing the net transfer of CO_2 from the atmosphere into water (Mara, 2005). During the daytime the Ginebra AFP attained maximum pH of 9.0 (**Figure 5.3**) – which can be explained by the CO_2 sink due to high-rate photosynthesis observed in this period. Similarly, in previous studies of constructed wetlands and natural systems, the negative fluxes have been explained by the high pH values (up to 9.0) measured in the water column (Søvik *et al.*, 2006). On the other hand, during night-time algae switch to respiration generating a near neutral pH, suggesting saturation of CO_2 in the water column, which leads to an increase of the mass transfer of this gas into the atmosphere (emission).

Tropical secondary algal facultative pond may serve as atmospheric sources of CO_2 because of higher local temperatures. Elevated temperatures stimulate decomposition of organic matter producing CO_2 (Stadmark and Leonardson, 2005; Detweiler *et al.*, 2014). CO_2 variations obtained in this study were higher than those reported in other waste stabilization ponds under subtropical and boreal conditions (Huttunen *et al.*, 2002; Hernandez-Paniagua *et al.*, 2014; Glaz *et al.*, 2016). Compared to temperate wastewater treatment systems, CO_2 fluxes were 8 and 20-fold lower than those in constructed wetlands (Liikanen *et al.*, 2006; Søvik *et al.*, 2006; Ström *et al.*, 2007). Likewise, the CO_2 consumption and emissions in the present study were 2 and 6-fold higher than those reported in shallow and deep lakes (Tremblay *et al.*, 2004; Xing *et al.*,

2005). The difference between tropical and temperate conditions is likely related to environmental condition – temperature being an obvious factor.

Table 5.6 Data of CO_2, CH_4, and N_2O fluxes from different wastewater treatments reported in the literature.

Location	Wastewater source	Measurement condition	CO_2[a] mg.m^{-2}.d^{-1}	CH_4[a] mg.m^{-2}.d^{-1}	N_2O[a] mg.m^{-2}.d^{-1}	References
Waste Stabilization Ponds						
Colombia	Municipal	Tropical	-3100-4730	80-4510	-6,3-15	Present study
Sweden	Municipal	Spring and summer	n.d.	-40-1700	-9-40	Johansson et al. (2003;2004)
India	Municipal	Subtropical	n.d.	n.d.	0.0-0.5	Singh et al. (2005)
Sweden	Municipal	Temperate	n.d.	11-970	n.d.	Stadmark and Leonardson (2005)
Mexico	Agricultural WSPs	Subtropical	200-1000	600±400	0.012-0.95	Hernandez-Paniagua et al. (2014)
USA	Municipal	Temperate	n.d.	3000-7400	n.d.	Detweiler et al. (2014)
Canada-Australia	Municipal	Temperate	-800-25700	0.7-608	n.d.	Glaz et al. (2016)
Duckweed Ponds						
USA	Synthetic stormwater	Temperate	1700-3300	500-1900	0.63-4	Sims et al. (2013)
USA	Synthetic stormwater	Temperate	0.4-1.4	200-600	n.d.	(Dai et al. 2015)

n.d. no data

Regardless of either COD or BOD, the organic mass contributes to CO_2 emissions in wastewater treatment systems (Teiter and Mander, 2005). This finding is corroborated by the positive correlation between COD and CO_2 fluxes in Ginebra's AFP (**Table 5.4**). Despite this, statistical analysis indicated that COD only accounts for 5.5% of CO_2 emissions in AFPs. Therefore, the mineralization of organic matter in the AFP may plausibly contribute to lowering the CO_2 exchange than algal photosynthesis or dark cycle respiration (Pulliam, 1993; Xing *et al.*, 2005).

5.4.2 CH₄ dynamics

CH_4 fluxes in the AFP of Ginebra were always positive. Indeed, under all conditions CH_4 was always emitted. Compared to other studies (**Table 5.6**), peak CH_4 emissions from AFPs were twofold higher than those of temperate ponds (Johansson *et al.* 2004), ten-times higher than subtropical ponds (Hernandez-Paniagua *et al.*, 2014) and nearly a hundredfold higher than boreal ponds (Glaz *et al.*, 2016). CH_4 emissions were comparable to those in a free water surface constructed wetland designed to purify peat mining runoffs (Liikanen *et al.*, 2006). However, FWS wetlands received a higher intake of organic load than AFPs. Notwithstanding other findings, CH_4 emissions in this study generally exceeded those wastewater treatments reported for AFPs in other studies (**Table 5.6**).

It seems high ambient temperatures ($>30^{\circ}C$) largely account for methane release from AFP. These findings mesh with previous studies in constructed wetlands where CH_4 release have been significantly higher during summer than during winter (Tanner *et al.*, 1997; Johansson *et al.*, 2004; Teiter and Mander, 2005; Xing *et al.*, 2005; Liikanen *et al.*, 2006; Picek *et al.*, 2007). Regression analysis supported the 38.7% CH_4 variation found in the AFP studied. According to Johansson *et al.* (2004) mineral nitrogen in the form of ammonium, nitrite, and nitrate reflected microbial activity in FWS-constructed wetlands. It is likely that similar mechanisms influenced the conditions for methane formation, consumption, and emission. Several reports indicated that NO_3^--N and NH_4^+-N had no effect on the CH_4 production in sediments from a pond and an eutrophic lake respectively (Liikanen and Martikainen, 2003; Stadmark and Leonardson, 2005). In contrast, Søvik and Klove (2007), found that CH_4 emissions were positively correlated to NO_3^--N and NH_4^+-N. The current study was unable to uncover any links between NTK, NO_3^--N, and NH_4^+-N and CH_4 production in the Ginebra's AFP. A possible reason was the minor variations in NTK, NO_3^--N, and NH_4^+-N in the water column which impacted the sensitivity of the regression model to fluctuations of these parameters.

In shallow secondary algal facultative ponds, algal photosynthesis augments oxygen concentrations in the water column, potentially inhibiting methanogenesis (Johansson *et al.*, 2004; Detweiler *et al.*, 2014). Using stable isotopes (Detweiler *et al.*, 2014) found an oxidation efficiency of 69.1%. Comparable numbers - 57%, 88%, 75% - were respectively recorded in lakes by (Bastviken *et al.*, 2002) , (Kankaala *et al.*, 2007), and (Schubert *et al.*, 2012). During daytime, the dissolved oxygen in the water column of the AFP reached oversaturation values of (>16 mg.l^{-1}) which seems to promote some methane oxidation effectively lowering methane emissions (**Figure 5.4**). However, in the absence of concrete data supporting methane oxidation efficiency, this hypothesis is tenuous and highlights the need for studies on mechanisms influencing methane oxidation in AFPs.

The dissolved methane from the anaerobic pond pre-treatment effluent may also have contributed to high emissions from the AFP. The average concentrations of CH_4 dissolved measured in the water column indicated that the AFP was super-saturated in CH_4. Similarly, maxima CH_4 saturation levels were recorded at dawn in AFPs (Glaz *et al.*, 2016). In addition to CH_4 produced under anaerobic conditions at the bottom of the AFP studied, this super-saturation of CH_4 also can be due to the anaerobic effluent being fed into the AFP. Thus, the dissolved and suspended micro-bubbles of CH4 from anaerobic effluents could lead to increased methane emissions from the AFP. This fact calls about recovering the methane dissolved from anaerobic effluents.

5.4.3 N_2O dynamics

N_2O variations in the Ginebra´s AFP were in the range reported from wastewater treatment plants (**Table 5.6**) (Johansson *et al.*, 2003; Singh *et al.*, 2005; Liikanen *et al.*, 2006; Søvik *et al.*, 2006; Søvik and Klove, 2007; Ström *et al.*, 2007). However, nutrient-rich wetlands consistently maintained higher N_2O emission levels than Ginebra´s AFP during late spring, early summer and autumn (Johansson *et al.*, 2003; Søvik and Klove, 2007; Ström *et al.*, 2007), which were observed during late spring, early summer and autumn. Negative N_2O fluxes suggest sequestration by the AFP and are consistent with the data measured in Sweden ponds (Johansson *et al.*, 2003; Ström *et al.*, 2007). Even lower than the AFP were the fluctuations from natural systems (Huttunen *et al.*, 2002) probably due to lower temperatures and nutrient levels in these bodies of water.

According to Wrage *et al.* (2001) two microbial types are involved in generating N_2O from wastewater: (i) Nitrifiers: ammonia-oxidizing bacteria (AOB) which produce N_2O mainly by incomplete oxidation of hydroxylamine and by nitrifier denitrification. In the latter, the oxidation of NH_3 to N_2O^- is followed by the reduction of NO_2^- to N_2O and N_2. In the case of nitrification, the N_2O is formed during incomplete NH_4^+-N oxidation to NO_3^-- N due to low DO concentration (Zeng *et al.*, 2003). ii) Denitrifiers produce N_2O as an intermediary during the metabolic reduction of NO_3^--N to N_2. Depending on nitrate availability, N_2O is formed by incomplete denitrification, increase of oxygen density in water and/or low COD/N ratio (3.5) (Hanaki *et al.*, 1992; Von Schulthess *et al.*, 1994; Søvik *et al.*, 2006).

Linear regression analysis applied to this study incorporated all components known to affect N_2O formation whether nitrification or denitrification. The inescapable inference is that N_2O dynamics were positively influenced by air-temperature, NO_3^--N and TKN, and negatively by DO and COD. These environmental parameters affecting the undulating cycles of N_2O have been reported elsewhere. In a constructed wetland with cultivated reed, the N_2O production correlated positively with NO_3^--N and soil temperature, whereas NH_4^+-N correlated negatively (Fey *et al.*, 1999). In a pond, nitrification was strongly linked to the formation ammonia, nitrite and total mineralized nitrogen, while the denitrification was related by nitrite, nitrate and total mineralized nitrogen (Johansson *et al.* 2003).

Oxygen released during algal photosynthesis influences daily N_2O emissions (**Figure 5.6**). In times of high solar radiation, AFPs achieve a peak DO concentration of 16.2 $mgO_2.L^{-1}$ Remarkably, a high DO concentration is conducive to nitrification – the driving force behind NO_3^-- N production. The corollary reinforces the observation: during the night-time, in the absence of photosynthesis, the average DO concentrations in the AFPs drops to 1.2±0.9 $mgO_2.L^{-1}$ thereby limiting nitrification. A likely proposition is that incomplete oxidation of hydroxylamine to NO_2^-N favors N_2O production. It can therefore be deduced that conditions favoring low DO is conducive to N_2O production by nitrifying organisms. Investigators using full-scale activated sludge showed that denitrification contributed between 58% and 83% of the N_2O emissions for oxygen concentrations around 1 $mgO_2.L^{-1}$ (Tallec *et al.*, 2006).

The inverse correlation observed for COD reinforces the hypothesis that N_2O was partially produced by incomplete denitrification. The operational COD/N ratio in the AFP studied was about 3.0, suggesting low availability of a carbon source in the AFP for the denitrification process. According to Itokawa *et al.* (2001), more than 20% of influent nitrogen can be emitted as N_2O during biological nitrogen removal in high strength wastewater when the COD/N ratio falls below 3.5. Von Schulthess *et al.* (1994) also found that the intensified N_2O emission from denitrification was fuelled by an increase in oxygen concentrations. In support, Oh and Silverstein (1999) noted that oxygen concentrations of 0.09 l^{-1} decreased denitrification by 35% and a total absence of denitrification at 5.9 mgl^{-1}. It is therefore reasonable to conclude that N_2O production in the Ginebra´s AFP is due to the presence of oxygen in the water column which in turn is transferable to anoxic sediments (Kampschreur *et al.*, 2009).

5.4.4 Global warming potential (GWP)

Based on mean emissions derived from this study, the GWP of CH_4 translates into 4.4 kg $CO_{2eq}.d^{-1}$, *w*hile for N_2O this is 0.03 kg $CO_{2eq}.d^{-1}$. Thus, the ratio CH_4/N_2O indicates that 99% of the GWP from Ginebra´s AFP were attributable to the release of methane gas. In addition, based on the total emissions of CH_4 and N_2O (4.43 kg $CO_{2eq}.d^{-1}$) and COD (1.5 $kgCOD_{rem}.d^{-1}$), the GWP of the AFP studied was 2.8 kg $CO_{2eq}.kgCOD_{removed}^{-1}$ or 20.8 kg CO_{2eq}. per p.e./year. According to the study, CO_{2eq} in the AFP are three times higher than that reported for activated sludge and aerobic/anaerobic hybrid reactor (1.04 kg $CO_{2eq}.kgCOD_{removed}^{-1}$ or 5.2 kg CO_{2eq}. per p.e./year) (Lundin *et al.*, 2000; Keller and Hartley, 2003; Bani Shahabadi *et al.*, 2009; Flores-Alsina *et al.*, 2011). Therefore, the AFP operated under the conditions described generated a higher GWP than conventional systems.

5.5 CONCLUSIONS

- This study generated new data on GHG dynamics from facultative algal ponds. Overall, AFPs are net suppliers of CH_4, CO_2, and N_2O. Among other environmental factors, the emissions of these gases are influenced by pH, ORP, N-NO_3, ambient temperature, DO and

COD. Moreover, daily variations also disclosed a significant role of photoperiod on GHG emissions on the AFP studied.

- Based on the data derived from the study, the compelling conclusion is that the AFPs emitted substantial volumes of GHG into the atmosphere compared to conventional systems. The largest contributor to GWP was CH_4. These findings provoke calls for additional studies to ensure better understanding of eco-technologies on enhanced sequestration capacity of wastewater treatment stabilization ponds and their sustainability regarding global warming.

5.6 ACKNOWLEDGEMENTS

This study has been carried out within the framework of the European research project SWITCH (Sustainable Urban Water Management Improves Tomorrow's City's Health). SWITCH is supported by the European Commission under the 6th Framework Programme and contributes to the thematic priority area of "Global Change and Ecosystems" [1.1.6.3] Contract n° 018530-2. The authors are also indebted to ACUAVALLE ESP and the Universidad del Valle for their support.

5.7 REFERENCES

APHA, AWWA and WEF (2005) Standard methods for the examination of water and wastewater, American Public Health Association, American Water Works Association and Water Environment Federation, Washington DC.

Arthur, J.P. (1983) Notes on the design and operation of waste stabilization ponds in warm climates of developing countries. Technical Paper 7, The World Bank, Washington, DC.

Bani Shahabadi, M., Yerushalmi, L. and Haghighat, F. (2009). Impact of process design on greenhouse gas (GHG) generation by wastewater treatment plants. Water research 43(10), 2679-2687.

Bastviken, D., Ejlertsson, J. and Tranvik, L. (2002). Measurement of methane oxidation in lakes: a comparison of methods. Environmental science & technology 36(15), 3354-3361.

Dai, J., Zhang, C., Lin, C.-H. and Hu, Z. (2015). Emission of Carbon Dioxide and Methane from Duckweed Ponds for Stormwater Treatment. Water Environment Research 87(9), 805-812.

Detweiler, A.M., Bebout, B.M., Frisbee, A.E., Kelley, C.A., Chanton, J.P. and Prufert-Bebout, L.E. (2014). Characterization of methane flux from photosynthetic oxidation ponds in a wastewater treatment plant. Water Science & Technology 70, 980-989.

Fey, A., Benckiser, G. and Ottow, J.C.G. (1999). Emissions of nitrous oxide from a constructed wetland using a groundfilter and macrophytes in waste-water purification of a dairy farm. Biology and fertility of soils 29(4), 354-359.

Flores-Alsina, X., Corominas, L., Snip, L. and Vanrolleghem, P.A. (2011). Including greenhouse gas emissions during benchmarking of wastewater treatment plant control strategies. Water research 45(16), 4700-4710.

Forster, P., Ramaswamy, V., Artaxo, P., Berntsen, T., Betts, R., Fahey, D.W., Haywood, J., Lean, J., Lowe, D.C., Myhre, G., Nganga, J., Prinn, R., Raga, G., Schulz, M. and Van Dorlan, R. (2007). Changes in atmospheric constituents and in radiative forcing. In: Climate Change 2007: The Physical Science Basis. . Solomon, S., D. Qin, M. Manning, Z. Chen, M. Marquis, K.B. Averyt, M.Tignor and H.L. Miller (ed), Cambridge University Press, Cambridge, United Kingdom and New York, NY, USA.

Glaz, P., Bartosiewicz, M., Laurion, I., Reichwaldt, E.S., Maranger, R. and Ghadouani, A. (2016). Greenhouse gas emissions from waste stabilisation ponds in Western Australia and Quebec (Canada). Water research 101, 64-74.

Hanaki, K., Hong, Z. and Matsuo, T. (1992). Production of nitrous oxide gas during denitrification of wastewater. Wat. Sci. Technol. 26(5/6), 1027-1036.

Hernandez-Paniagua, I.Y., Ramirez-Vargas, R., Ramos-Gomez, M.S., Dendooven, L., Avelar-Gonzalez, F.J. and Thalasso, F. (2014). Greenhouse gas emissions from stabilization ponds in subtropical climate. Environmental Technology 35, 727-734.

Huttunen, J.T., Väisänen, T.S., Heikkinen, M., Hellsten, S., Nykänen, H., Nenonen, O. and Martikainen, P.J. (2002). Exchange of CO_2, CH_4 and N_2O between the atmosphere and two northern boreal ponds with catchments dominated by peatlands or forests. 137-146.

Itokawa, H., Hanaki, K. and Matsuo, T. (2001). Nitrous oxide production in high-loading biological nitrogen removal process under low COD/N ratio condition. Water research 35(3), 657-664.

Johansson, A.E., Gustavsson, M., Oquist, M.G. and Svensson, B.H. (2004). Methane emissions from a constructed wetland treating wastewater--seasonal and spatial distribution and dependence on edaphic factors. Water research 38(18), 3960-3970.

Johansson, A.E., Klemedtsson, A.K., Klemedtsson, L. and Svensson, B.H. (2003). Nitrous oxide exchanges with the atmosphere of a constructed wetland treating wastewater. Parameters and implications for emission factors. Tellus B 55(3), 737-750.

Kampschreur, M.J., Temmink, H., Kleerebezem, R., Jetten, M.S.M. and van Loosdrecht, M.C.M. (2009). Nitrous oxide emission during wastewater treatment. Water research 43(17), 4093-4103.

Kankaala, P., Eller, G. and Jones, R.I. (2007). Could bacterivorous zooplankton affect lake pelagic methanotrophic activity? Fundamental and Applied Limnology/Archiv für Hydrobiologie 169(3), 203-209.

Keller, J. and Hartley, K. (2003). Greenhouse gas production in wastewater treatment: process selection is the major factor. Water Science and Technology 47(12), 43-48.

Liikanen, A., Huttunen, J.T., Karjalainen, S.M., Heikkinen, K., Vaisanen, T.S., Nykanen, H. and Martikainen, P.J. (2006). Temporal and seasonal changes in greenhouse gas emissions from a constructed wetland purifying peat mining runoff waters. Ecological Engineering 26(3), 241-251.

Liikanen, A. and Martikainen, P.J. (2003). Effect of ammonium and oxygen on methane and nitrous oxide fluxes across sediment-water interface in a eutrophic lake. Chemosphere 52(8), 1287-1293.

Lundin, M., Bengtsson, M. and Molander, S. (2000). Life cycle assessment of wastewater systems: influence of system boundaries and scale on calculated environmental loads. Environmental science and technology 34(1), 180-186.

Mara, D. (2005). Pond treatment technology. Shilton, A. (ed), pp. 168-187, IWA Publishing London.

Metcalf, L. and Eddy, H. (2003) Wastewater Engineering: Treatment, Disposal and Reuse, , Mc Graw Hill, New York.

Oh, J. and Silverstein, J.A. (1999). Oxygen inhibition of activated sludge denitrification. Water research 33(8), 1925-1937.

Peña, M.R., Madera, C.A. and Mara, D.D. (2002). Feasibility of waste stabilization pond technology for small municipalities in Colombia. Water Science and Technology 45(1), 1-8.

Picek, T., Cizkova, H. and Dusek, J. (2007). Greenhouse gas emissions from a constructed wetland--Plants as important sources of carbon. Ecological Engineering 31(2), 98-106.

Pulliam, W.M. (1993). Carbon dioxide and methane exports from a southeastern floodplain swamp. Ecological Monographs, 29-53.

Schubert, C.J., Diem, T. and Eugster, W. (2012). Methane emissions from a small wind shielded lake determined by eddy covariance, flux chambers, anchored funnels, and boundary model calculations: a comparison. Environmental science & technology 46(8), 4515-4522.

Shilton, A., Mara, D., Craggs, R. and Powell, N. (2008). Solar-powered aeration and disinfection, anaerobic co-digestion, biological CO2 scrubbing and biofuel production: the energy and carbon management opportunities of waste stabilisation ponds. Water Science and Technology 58(1), 253-258.

Silva, J.P., Lasso, A., Lubberding, H.J., Peña, M.R. and Gijzen, H.J. (2015). Biases in greenhouse gases static chambers measurements in stabilization ponds: Comparison of flux estimation using linear and non-linear models. Atmospheric Environment 109, 130-138.

Silva, J.P., Ruiz, J.L., Peña, M.R., Lubberding, H. and Gijzen, H. (2012). Influence of photoperiod on carbon dioxide and methane emissions from two pilot-scale stabilization ponds. Water Science and Technology 66(9), 1930-1940.

Sims, A., Gajaraj, S. and Hu, Z. (2013). Nutrient removal and greenhouse gas emissions in duckweed treatment ponds. Water research 47(3), 1390-1398.

Singh, V.P., Dass, P., Kaur, K., Billore, S.K., Gupta, P.K. and Parashar, D.C. (2005). Nitrous oxide fluxes in a tropical shallow urban pond under influencing factors. Current Science 88(3), 478.

Søvik, A.K., Augustin, J., Heikkinen, K., Huttunen, J.T., Necki, J.M., Karjalainen, S.M., Klove, B., Liikanen, A., Mander, U. and Puustinen, M. (2006). Emission of the Greenhouse Gases Nitrous Oxide and Methane from Constructed Wetlands in Europe. Journal of Environmental Quality 35(6), 2360.

Søvik, A.K. and Klove, B. (2007). Emission of N_2O and CH_4 from a constructed wetland in southeastern Norway. Science of the Total Environment 380(1-3), 28-37.

Stadmark, J. and Leonardson, L. (2005). Emissions of greenhouse gases from ponds constructed for nitrogen removal. Ecological Engineering 25(5), 542-551.

Ström, L., Lamppa, A. and Christensen, T.R. (2007). Greenhouse gas emissions from a constructed wetland in southern Sweden. Wetlands Ecology and Management 15(1), 43-50.

Tallec, G., Garnier, J., Billen, G. and Gousailles, M. (2006). Nitrous oxide emissions from secondary activated sludge in nitrifying conditions of urban wastewater treatment plants: effect of oxygenation level. Water research 40(15), 2972-2980.

Tanner, C.C., Adams, D.D. and Downes, M.T. (1997). Methane emissions from constructed wetlands treating agricultural wastewaters. Journal of Environmental Quality 26(4), 1056-1062.

Teiter, S. and Mander, U. (2005). Emission of N_2O, N_2, CH_4, and CO_2 from constructed wetlands for wastewater treatment and from riparian buffer zones. Ecological Engineering 25(5), 528-541.

Toprak, H. (1995). Temperature and organic loading dependency of methane and carbon dioxide emission rates of a full-scale anaerobic waste stabilization pond. Water research 29(4), 1111-1119.

Tremblay, A., Lambert, M. and Gagnon, L. (2004). Do hydroelectric reservoirs emit greenhouse gases? Environmental Management 33, 509-517.

Van der Steen, N.P., Nakiboneka, P., Mangalika, L., Ferrer, A.V. and Gijzen, H.J. (2003). Effect of duckweed cover on greenhouse gas emissions and odour release from waste stabilisation ponds. Water Sci Technol 48(2), 341-348.

Von Schulthess, R., Wild, D. and Gujer, W. (1994). Nitric and nitrous oxides from denitrifying activated sludge at low oxygen concentration. Water Science and Technology 30(6), 123-132.

Wrage, N., Velthof, G., Van Beusichem, M. and Oenema, O. (2001). Role of nitrifier denitrification in the production of nitrous oxide. Soil Biology and Biochemistry 33(12), 1723-1732.

Xing, Y., Xie, P., Yang, H., Ni, L., Wang, Y. and Rong, K. (2005). Methane and carbon dioxide fluxes from a shallow hypereutrophic subtropical Lake in China. Atmospheric Environment 39(30), 5532-5540.

Zeng, R.J., Lemaire, R., Yuan, Z. and Keller, J. (2003). Simultaneous nitrification, denitrification, and phosphorus removal in a lab-scale sequencing batch reactor. Biotechnology and bioengineering 84(2), 170-178.

Chapter 6

Anthropogenic Influence on Greenhouse Gas fluxes in a Tropical Natural Wetland

Abstract

This study determined the fluxes of CO_2, CH_4 and N_2O emitted from a eutrophic tropical freshwater wetland (FW) called Sonso Lagoon. This FW receives pollution from several sources including agricultural run-off, domestic sewage, and a polluted river. The results indicate that the fluxes for CO_2, CH_4 and N_2O showed a large variation ranging from consumption to emissions. CO_2 fluxes ranged from -22.9 to 23 $g.m^{-2}.d^{-1}$ (median = 0.93), CH_4 ranged between -3.03 and 9.83 $g.m^{-2}.d^{-1}$ (median = 0.04), and N_2O ranged from -15.2 to 12.6 mg N_2O $m^{-2}.d^{-1}$ (median = 0.21). For all the three gases studied, negative fluxes were observed mainly in the zone dominated by floating aquatic macrophytes i.e. *Eichornia crassipes, Salvinia sp.*, and *Pistia stratiotes* L. However, the mean values indicated that the Sonso Lagoon was a net source of GHG production. The effect of eutrophication on GHG emissions could be observed in the positive correlation found between CH_4 and CO_2 generation and COD, PO_4^{-3} and NH_4-N. In addition, N_2O correlated positively to TKN and NO_3^-N. This study demonstrates that pollution and eutrophication of natural wetlands results in net emissions of greenhouse gases into the atmosphere.

Key words: Freshwater wetland, Greenhouse Gases, Eutrophication, Floating Aquatic Macrophytes, Methane fluxes, Nitrous oxide fluxes, carbon dioxide fluxes

6.1 INTRODUCTION

Freshwater Wetlands (FWs) provide ecosystem and socio-economic services such as water purification, flood control, nutrient cycling, carbon sequestration, foods i.e. aquaculture, maintenance of biodiversity and climate regulation (Zedler and Kercher, 2005; Mitsch and Gosselink, 2007). However, freshwater wetlands can also be an important source of greenhouse gases (GHGs). The main natural sources of CH_4 are wetlands (177 to 284 $TgCH_4.year^{-1}$). It has been estimated that CH_4 emissions from tropical wetlands are 128 $Tg.yr^{-1}$ and equivalent to 75% of the total emissions from wetlands worldwide (Anderson *et al.*, 2010). Wetlands have been neglected as a source of N_2O; it was determined that arctic wetlands may emit 32,400-340,200 metric tons of CO_2 equivalents that is approximately 0.01 and 1% of total GHG emissions for 2005 (39 billion metric tons of CO_2 equivalent) (Ventura, 2014). Further, recent research suggests that the drainage of wetlands, especially peatlands, may increase emissions up to 2 Tg N_2O-N. yr^{-1}. The global CO_2 emission rates are 1800 Tg $C.yr^1$ from streams and rivers and 320 Tg C. yr^{-1} from lakes (Raymond *et al.*, 2013).

These emissions can be even higher where wetlands have been affected by anthropogenic activities such as agricultural run-off and domestic sewage discharges. These activities increase nutrient input (N and P) and organic matter into FWs, resulting in severe alterations of the water quality and function of these ecosystems i.e. eutrophication (Ventura, 2014). Eutrophication in FW generally promotes excessive algae and plant growth and decay, favouring simple algae and plankton over other more complicated plants, and this can cause severe reduction in water quality as a result of oxygen depletion (Søndergaard, 2007). In addition, the nutrient loading enhances organic matter decomposition and microbial activity (Wright et al., 2009), which may lead to increased accumulation of carbon and nitrogen. As a result, eutrophication affects the freshwater wetland biogeochemistry, leading to acceleration of the exchange of greenhouse gases between freshwater wetlands and the atmosphere (Casper et al., 2000; Huttunen et al., 2001).

In wetlands, CO_2 emissions result from the decomposition of organic matter (OM) present in the aquatic bodies, including OM from forests, soils, vegetation, upstream rivers and photosynthetic fixation by phytoplankton on the surface (Yang et al., 2014). CO_2 is produced by respiration in the sediments and in the water column, as well as by other biological processes in the sediment. This CO_2 produced may be stored near the sediment / water interface and then transferred into the atmosphere due to super-saturation of CO_2 in the wetland (Schrier-Uijl et al., 2011). CO_2 emissions in wetlands generally occur in two ways: diffusion and boiling. Diffusion is the main process for the emission of this gas, whereas the contribution by boiling is lower because of the high solubility of CO_2, which is easily absorbed by water (Yang et al., 2014).

CH_4 emissions are the result of the balance of two processes: methanogenesis in anoxic conditions and oxidation of the generated methane (Minkkinen and Laine, 2006). CH_4 production in FW originates from microbial metabolism by methanogenic archaea when the organic matter is decomposed in the absence of oxygen or of other electron acceptors such as: NO_3^- (denitrification), Fe (III) (iron reduction), Mn (III, IV) (manganese reduction) and $SO4^{-2}$ (sulphate reduction) (Bridgham et al., 2013). CH_4 emission from wetlands is mainly due to processes of boiling and molecular diffusion. The produced CH_4 moves from the sediment through the water column into the atmosphere; however, along this trajectory, it can be oxidized to CO_2. Non-oxidized CH_4 can be emitted into the atmosphere by diffusion (Schrier-Uijl et al., 2011; Yang et al., 2014).

Another GHG produced in this type of aquatic ecosystems is N_2O, which is produced in aerobic and anaerobic conditions, depending on the substrate and microorganisms involved. NO_3^- can be reduced to NO_2 and N_2 via microbial denitrification in anaerobic conditions (Beaulieu et al., 2011). N_2O is produced as an intermediate sub-product in the transformation of NO_3^- to N_2, whereas in aerobic conditions, NH_4^+ can be oxidized to N_2O and NO_3^- by nitrifying bacteria (Hendzel et al., 2005). The main process by which N_2O is emitted into the atmosphere is the diffusion process because of the low solubility of this gas (Yang et al., 2014).

At the global level, scientific research has focused on the contribution of wetlands to GHG emissions for wetlands in boreal climate (Huttunen et al., 2003; Liikanen et al., 2003; Song et al., 2012), arid climate (Duan et al., 2005), subtropical climate (Wang et al., 2007) and temperate climate (Schrier-Uijl et al., 2011). However, studies on the dynamics and factors that influence GHG emissions in natural tropical climate wetlands are limited (Mitsch *et al.*, 2013). Yet, wetlands could provide low cost eco-technologies for wastewater treatment especially for developing countries, which are mostly located in tropical environment. Thus, it is important to study to what extent GHG emissions occur from wetlands under tropical conditions.

In this regard, the objectives of this research are (i) to quantify GHG emissions in a tropical freshwater wetland in Colombia; (ii) to determine the spatial variation of the GHG emissions; and (iii) to identify factors that influence the emissions of CH_4, CO_2 and N_2O from a eutrophic tropical freshwater wetland, such as aquatic plant cover, temperature, pH and nitrate concentrations.

6.2 METHODOLOGY

6.2.1 Site description
The Sonso Lagoon (3°51′43.36″ N and 76°20′57.12″ W) is a shallow eutrophic lake. It is located in the southwestern part of Colombia, on the right bank of the Cauca River (**Figure 6.1**)..

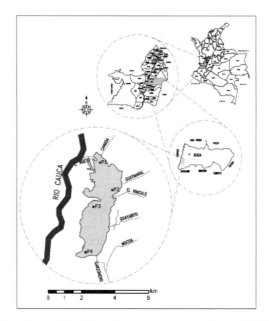

Figure 6.1 Scheme location of the Sonso Lagoon and sampling points for measurements of greenhouse gas emissions

The total area of the Sonso Lagoon is 20.5 km^2 of which 7.5 km^2 are water mirror and 13 km^2 riparian buffer zones. It has a mean depth of 1.6 m with a maximum depth of 3 m. The climate is tropical with a mean annual temperature ranging from 21 to 26°C and 1,375 mm annual precipitation.

The Sonso Lagoon exchanges water and sediments with the Cauca River through the Caño Nuevo Channel (**Figure 6.1**). In addition, there are discharges of domestic sewage and agricultural runoff from sugar cane crops into the lake. Organic matter, nitrogen compounds, phosphorus compounds, heavy metals, pesticides and herbicides enter the lagoon through these discharges, affecting water quality. Due to eutrophication, the lagoon is covered with water plants, i.e. hyacinths.

The four study sites in the Sonso Lagoon were as follows (**Figure 6.1**): P1, exchange zone of water and sediments with the Cauca River; P2, zone dominated by domestic sewage and agricultural runoff discharges; P3, dominated by phytoplankton; P4, water-floating plant-covered areas (mostly water hyacinth).

6.2.2 Water and sediment characteristics
Environmental data from the four sites of the Sonso Lagoon were collected to determine the water quality and sediments to evaluate the relationship between greenhouse gases and environmental parameters. The sampling and measurement of environmental data were carried out between 9 am and 3 pm, which coincided with the greenhouse gas emission measurements. Chemical oxygen demand (COD), total suspended solids (TSS), alkalinity, total Kjeldahl nitrogen (TKN), ammonium nitrogen (NH_4^+-N), and nitrate nitrogen (NO_3^--N) were measured according to Standard Methods (APHA, 2005). Conductivity, pH, dissolved oxygen (DO), temperature and oxidation-reduction potential were measured with electrodes at a depth of 30 cm.

6.2.3 GHG measurements
Emissions of CH_4, CO_2, and N_2O were measured with the static chamber technique at the four sites defined above (Silva *et al.*, 2015). The static chambers were of propylene and cylindrical (diameter 0.43 m and height 0.26 m). The chambers were equipped with a sampling port having a rubber septum from which the gas sample was taken. During measurements, the chambers were installed gently on the water surface of each point. Samples of gas (20 mL) for CH_4 CO_2 and N_2O measurements were taken during 30 minutes at 10-minute intervals (0, 10, 20, 30 min) from a sampling port on the top of each chamber using a special syringe. Finally, the gas samples were withdrawn directly through a needle into pre-evacuated containers of 10 mL volume.

Methane was analysed by gas chromatography (Shimadzu Co., Japan) equipped with a flame ionic detector (FID) and a Porapak Q column, and the temperature of the oven and injector ports were set at 60 and 80°C, respectively. CO_2 was measured by an infrared spectrophotometer Qubit S151 CO_2 analyser (Loligo Systems, Denmark) using 75 ml min^{-1} air as the mobile phase with the temperature of the injector set equal to the ambient temperature. The N_2O concentration was analysed by means of gas chromatography (Shimadzu Co., Japan) equipped with an electron capture detector (ECD) and a Porapak column Q 80–100 mesh 2 m*2 mm retention gap, using 22 ml min^{-1}. N_2 was the carrier gas, and the injector, column, and detector temperatures were 80, 70, and 320 °C, respectively.

The flux of CH_4, CO_2 and N_2O were calculated according to Equation 1 from linear and non-linear changes in the gas concentrations in the chamber headspaces (Silva et al., 2015).

$$F = \frac{dC}{dt_{t=0}} \times \frac{V_c}{A} \times \frac{1440\ min}{d} \quad \text{(Eq. 6.1)}$$

F = Flux of CH_4, CO_2 y N_2O (g.m^{-2}.d^{-1}); dC/dt $_{t=0}$ = Slope of the gas concentration curve (g.m^{-3}.d^{-1}); V_c = Volume of the chamber (m^3); A_c = The cross-sectional area of the chamber (m^2).

6.2.4 Data analysis
Statistical analyses were done with SPSS® software (V.17.0 for Windows). The Kolmogorov-Smirnov test was used to check the normality of the data. This normality check was done to determine whether parametric or non-parametric tests should be applied to analyse the data. To determine the differences in GHG fluxes from the different sample points, the Kruskal-Wallis and Mann Whitney non-parametric tests ($\alpha = 0.05$) were used. In addition, the relationship between the environmental data and GHG fluxes was estimated by using Spearman correlation coefficients.

6.3 RESULTS

6.3.1 Water characteristics
Table 6.1 shows the concentrations of dissolved oxygen (DO), total phosphorus (TP), and total nitrogen (TN) measured in the Sonso Lagoon. The DO mean values for the four points were ranging between 4.4± 2.2(P1) and 5.4 ± 1.6 (P3) mg.l^{-1}. This suggests that in general there was oxygen depletion in the water column and Sonso Lagoon was under-saturation DO condition. The low DO concentrations and excessive nutrient input provided the accumulation of reduced compounds such as ferrous iron and the release of ammonium and orthophosphates from the sediments into the water column, which triggered eutrophication (Boström et al., 1988; Ahlgren et al., 2011; Bellido et al., 2011).

COD concentrations in water observed between 26.8± 2.5 and 30.4 ± 2.6 mgO$_2$L^{-1} suggest moderate OM contamination in the Sonso Lagoon (Deborah, 1996). In addition, the data for TN (1.2 to 1.8 mg.l^{-1}) and TP (0.05 and 0.08 mg l^{-1}) indicated that the eutrophic state dominates (Nicholls et al., 2007; Smith and Smith, 2007; Serediak, 2014). Considering the Carlson Trophic State Index (TSI), it was found that the trophic state condition of the wetland varies from mesotrophic to eutrophic; thus, points P1 and P2 were in the mesotrophic state, and points P3 and P4 were in the eutrophic state. The higher TSI index at P3 and P4 can be explained because the direct discharge of domestic wastewater from the population adjacent to this zone of Sonso Lagoon which in turn also causes an increase in nutrients (N and P) concentrations.

Table 6.1 Water characteristics of the Sonso Lagoon. (mean ± s.d.)

Parameter	Sampling Point			
	P1	**P2**	**P3**	**P4**
Depth (m)	0.6 ± 0.3	0.5 ± 0.5	0.7 ± 0.5	0.6 ± 0.6
Water temperature (°C)	26.8 ± 2.5	28.7 ± 2.9	30.3 ± 2.8	30.4 ± 2.6
Water transparency (m)	0.1 ± 0.1	0.1 ± 0.1	0.1 ± 0.1	0.2 ± 0.1
pH (units)	7.8 ± 0.5	7.8 ± 0.4	7.9 ± 0.4	7.8 ± 0.5
Alkalinity (mg.l^{-1} CaCO$_3$)	172.8 ± 52.2	215.7 ± 55	227.2 ± 37	256.2 ± 40
DO (mg. l^{-1})	4.4 ± 2.2	4.5 ± 1.7	5.4 ± 1.6	4.6 ± 1.6
COD (mg O$_2$.l^{-1})	33.8 ± 9.5	31.6 ± 12	31.7 ± 12.3	32.1 ± 12.2
TN (mg.l^{-1})	1.3 ± 1.1	1.2 ± 1.4	1.8 ± 2.1	1.7 ± 1.8
NH$_4^+$-N (mg.l^{-1})	1.0 ± 0.52	0.9 ± 0.64	0.7 ± 0.42	0.5 ± 0.3
NO$_2^-$- N (mg.l^{-1})	0.008 ± 0.007	0.007 ± 0.005	0.007 ± 0.006	0.004 ± 0.005
NO$_3^-$-N (mg.l^{-1})	0.06 ± 0.08	0.08 ± 0.11	0.07 ± 0.09	0.12 ± 0.12
TP (mg .L^{-1})	0.17 ± 0.11	0.31 ± 0.15	0.24 ± 0.08	0.08 ± 0.06
PO$_4^{-3}$- P (mg.l^{-1})	0.07 ± 0.04	0.16 ± 0.09	0.11 ± 0.07	0.05 ± 0.08

6.3.2 Greenhouse gas fluxes

Methane and carbon dioxide

The median methane fluxes for P1, P2, P3 and P4 were 0.7, 0.9, 0.2, and -0.4 g.m^{-2}.d^{-1}, respectively (**Figure 6.2**). This indicates that P1, P2 and P3 were net sources of methane whereas P4 in general was a sink. The only significant difference in flux was observed between P1 and P4 (p = 0.012).

Table 6.2 shows the correlations of Pearson between CH$_4$ fluxes and water characteristics at the four points. No correlation could be identified between methane flux and water temperature or the ORP (p>0.05). Positive correlations (p<0.05) were found between methane flux, PO4^{-3} and COD at P1. By contrast, in P2, DO and pH showed negative correlation whereas COD, NO$_3^-$-N, NH$_4^+$-N, were correlated positively to the CH$_4$ emissions. Although in P3 and P4 there were correlations between methane fluxes and PO4^{-3} (r = 0.57; p = 0.112), and NO$_3$-N (r = -.68; p = 0.45), it was not significant (p>0.05).

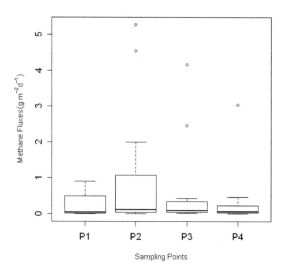

Figure 6.2 Fluxes of CH4 in the Sonso Lagoon. P1, P2, P3 are the open water sampling points whereas P4 is a hyacinth-covered area.

Table 6.2 Pearson correlations between CH_4 fluxes and water characteristics

Parameter	P1	P2	P3	P4
DO	n.o.	(0.048; -50%)	n.o.	n.o.
COD	(0.03; 55%)	(0.03; 62%)	n.o.	n.o.
PO_4^{-3}- P	(0.04; 66%)	n.o	(0.112; 57%)	n.o.
NO_3^--N	n.o.	(0.044; 59%)	(0.12; 78%)	(0.45; -68%)
NH_4^+-N	n.o.	(0.01; 77%)	n.o.	n.o.
pH	n.o.	(0.001; -75%)	n.o.	n.o.

n.o. = Pearson correlation was not observed

The mean CO_2 fluxes for P1, P2, P3, and P4 were 4.2, 3.9, 3.1 and -5.5 $g.CO_2 m^{-2}$ d^{-1}, respectively (**Figure 6.3**). The statistic comparison between the four points indicated that P4 showed significant differences compared to P1, P2, and P3 (P1: $p = 0.0002$; P2: $p = 0.0001$; P3: $p = 0.0001$; Mann Witney test). P4 seems to be a CO2 sink contributing to carbon sequestration in the freshwater wetland (**Figure 6.3**).

COD, TP, NO_3^--N, and NH_4^+-N were the most important parameters influencing CO_2 fluxes from the Sonso Lagoon) (**Table 6.3**). At P1, CO_2 fluxes were positively correlated to COD and negatively correlated to NH_4^+-N ($p<0.05$). In P2, the concentration measured of TP and NO_3^--N ($r = 0.89$; $p = 0.06$) showed a positive correlation with CO_2 fluxes ($p>0.05$). In P2 a negative correlation between NH_4^+-N and CO_2 release ($r = -.55$; $p = 0.04$) was also observed. P3 showed no significant correlation between CO_2 fluxes and the environmental parameters. In P4, the CO_2 fluxes were negatively correlated to the alkalinity ($r = -.71$; $p = 0.05$).

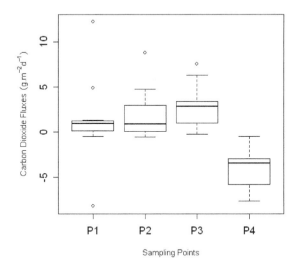

Figure 6.3 Fluxes of CO_2 in the Sonso Lagoon. P1, P2, P3 are the open water sampling points whereas P4 is a hyacinth-covered area

Table 6.3 Pearson correlations between CO_2 fluxes and water characteristics

Parameter	P1	P2	P3	P4
COD	(0.05; 57%)	n.o.	n.o.	n.o.
TP	n.o.	(0.001; 0.58)	n.o.	n.o.
NO_3^--N	n.o.	(0.06; 89%)	n.o.	(0.45; -68%)
NH_4^+-N	(-.62; p = 0.04).	(r = -.55; p = 0.04)	n.o.	n.o.

n.o. = Pearson correlation was not observed

Nitrous oxide

The median N_2O fluxes for P1, P2, P3, and P4 were 0.13, 0.3, 0.63 and -2.6 mg N_2O m^{-2} d^{-1}, respectively (**Fig. 6.4**). The negative median fluxes in P4 indicate that this point acts as a N_2O sink. At the other points N_2O was emitted. Only P4 showed significant differences compared to P1, P2, and P3. According to the Pearson correlation, at P1 the production of N_2O was correlated positively to TN ($r = 0.69$; $p = 0.026$) and NO_3^- -N ($r = 0.6$; $p = 0.09$). P2, P3, while P4 showed no correlation between environmental parameters and N_2O fluxes.

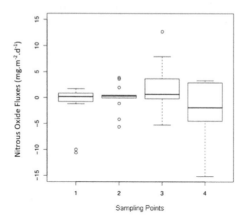

Figure 6.4 Flux of N_2O in the Sonso Lagoon. P1, P2, P3 are the open water sampling points whereas P4 is a hyacinth-covered area

6.4 DISCUSSION

6.4.1 CH₄ and CO₂ dynamics

For the entire wetland and considering only the average results of open water fluxes (P1, P2, P3; i.e. excluding macrophyte covered areas – P4), the Sonso Lagoon was a major source of CO_2 and CH_4 emission to the atmosphere. The CO_2 emissions were bigger than observed in other natural systems (Huttunen *et al.*, 2002; Tremblay *et al.*, 2004; Xing *et al.*, 2005; Schrier-Uijl *et al.*, 2011), but are comparable to values reported from constructed wetlands treating municipal wastewater (Liikanen *et al.*, 2006; Ström *et al.*, 2007). Similarly, the methane emissions from the Sonso Lagoon were mostly higher than those observed in other freshwater wetlands (Huttunen *et al.*, 2002; Tremblay *et al.*, 2004; Xing *et al.*, 2005; Schrier-Uijl *et al.*, 2011) and comparable to eco-technology systems for wastewater treatment (Johansson *et al.*, 2004; Liikanen *et al.*, 2006; Ström *et al.*, 2007; Silva *et al.*, 2012).

These differences can be attributed to the relatively high pollution of the wetland by nutrients from the River Cauca, agricultural run-off, and domestic sewage discharges leading to eutrophication and organic matter build-up. In the case of non-anthropogenic intervention, most freshwater wetlands are net carbon sinks because primary production is often able to sequester carbon at higher rates than biodegradation of organic matter into freshwater lake sediments over time (Brix *et al.*, 2001). However, in natural wetlands that receive wastewater i.e. the Sonso Lagoon, this ability to sequester carbon is affected because the nutrient and organic-matter loading seems to favour biodegradation leading to CH_4 and CO_2 emissions (Brix *et al.*, 2001; Huttunen *et al.*, 2003; Liikanen *et al.*, 2006; Wright *et al.*, 2009; Ventura, 2014).

CH_4 and CO_2 emissions were lower in the areas covered by floating vegetation such as *Eichhornia crassipes, Salvinia sp.*, and *Pistia stratiotes* (P4) than observed in open water areas (P1, P2 and P3). Although in the current study the greenhouse gas fluxes were measured only during the photoperiod, and macrophytes abundance was not quantified, it seems that macrophytes can help regulate C emissions in wetlands. These results are consistent with previous reports of C uptake by floating macrophytes (Attermeyer *et al.*, 2016). Reports show that net CO_2 flux can be three times higher at locations without macrophytes than where there is plant CO_2 uptake (Bolpagni *et al.*, 2007). Xing *et al.*, (2006) showed that macrophytes, rather than phytoplankton, directly positively affected net C emissions. In a tropical floodplain lake ecosystem, a macrophytes cover of around 40-50% could substantially offset open water CO_2 emissions and the lake would start to be a CO_2 sink (Peixoto *et al.*, 2016). In the Everglades it has also been suggested that the floating macrophytes contribute more to C uptake per area than phytoplankton in the open water (Schedlbauer *et al.*, 2012). Thus, the CO_2 and CH_4 uptake data found in the Sonso Lagoon suggests that the macrophytes-covered water can be important for carbon cycling in tropical wetlands.

The capacity of floating macrophytes to sequester CH_4 and CO_2 emissions are regulated by vegetation through: (i) acting as a physical barrier to prevent C diffusion across the water interface into the atmosphere (Van der Steen *et al.*, 2003; Silva *et al.*, 2012); (ii) sequestering CO_2 by algae and aquatic-floating plant photosynthesis (Brix *et al.* 2001; Teiter and Mander 2005; Ström *et al.* 2007); (iii) favouring methane oxidation through translocation of oxygen gas produced by photosynthetic activity of the green leaves to the stems and roots and to the water body (Laanbroek, 2010) and (iv) the presence of attached methylotrophs in biofilms attached to floating leaves (Whalen, 2005; Chowdhury and Dick, 2013; Wang *et al.*, 2013). However, these processes were not quantified in this study and therefore this will be a subject for further studies to elucidate the extension of these mechanisms on GHG emissions from wetlands.

Another factor that influences the gas transfer between sediment and air is the macrophytes rooting in sediments. Higher C emissions were observed in the presence of emergent macrophytes compared to floating vegetation in an arid lake area in western China (Duan *et al.*, 2005) and a southern boreal lake (Bergström *et al.*, 2007). The higher gas-releasing capacity in emergent macrophytes may be related to their rooting in sediments and continuous access to the atmosphere that provides a high potential to exchange gases primarily by internal pressurization and convective gas flows (Brix *et al.*, 1996; Sorrell and Brix ,2003; Laanbroek, 2010). However, floating plants like *Eichhornia crassipes, Salvinia sp.*, and *Pistia stratiotes*, found in the Sonso Lagoon, do not root in sediments and therefore their gas exchange capacity is lower than emergent macrophytes. This could also explain the low GHG emissions from P4 observed in this study.

The variability of CO_2 and CH_4 fluxes in the Sonso Lagoon could not be explained by changes in water temperature. This is because all the data were collected during daytime during maximum solar radiation and no significant changes in the water temperature were observed for the different sampling/monitoring campaigns. The studies in boreal and temperate conditions have reported the influence of water temperature on GHG emissions and C storage (Huttunen et al., 2003; Johansson et al., 2004; Xing et al., 2005). Changes of temperature in these wetlands between seasons affect   primary productivity influencing the pattern of production of CO_2, CH_4, and N_2O (Davidson et al., 2015).. However, in tropical wetlands these differences are hard to find because these ecosystems are exposed to a rather constant high mean annual temperature with little seasonal variation

Carbon dioxide emissions from the majority of the points in the Sonso Lagoon (P1, P2, P3) were correlated to NO_3^-, NH_4^+, and TP concentrations. The presence of these nutrients introduces shifts in decomposition rates and nutrient cycling mainly when wetlands become eutrophic. The positive correlation of NH_4^+ with CO_2 emission is in line with the findings of wetlands in The Netherlands (Schrier-Uijl et al., 2011). Nitrates serve as the first terminal electron acceptor in wetland soils after oxygen depletion, making them an important compound in the oxidation of organic matter in wetlands (Sánchez-Carrillo et al., 2011). Denitrifying bacteria play an important role in the carbon cycle of wetlands as they contribute significantly to the carbon mineralization budget (up to 50%) in eutrophic freshwaters (Christensen et al., 1990). The positive correlation between TP and CO_2 emissions can be explained by the fact that bacteria depend on phosphorous as a nutrient in microbial organic matter decomposition (Wright et al., 2008).

The fluxes of CH_4 correlated negatively with both pH and DO, and positively with COD at point P1 and P2. Comparing P1 and P2 to P3 and P4, it can be observed that in addition to pH, DO and COD other factors such as PO_4^{-3} and NO_3^- are regulating methane production at P3 and P4. In addition, as mentioned previously, the floating plants regulated methane emissions at P4.

The negative correlation between pH and CH_4 fluxes can be explained by the optimum conditions favouring methanogenesis, which is between pH 6.0 and 7.0. In this study with pH values higher than 7.0, CH_4 production might have been reduced by high pH. A negative correlation between pH and CH_4 also was reported from lakes and drainage ditches in temperate wetlands (Schrier-Uijl et al., 2011)

The statistical test indicated that increasing DO concentrations in the wetland methane emission decreased. In wetlands, DO concentration may vary dynamically due to the photosynthetic activity of plants and phytoplankton, affecting methane production (Harrison et al., 2005). The DO concentrations measured in Sonso Lagoon were ranging between 4.4 ± 2.2 and 5.4 ± 1.6 $mg.l^{-1}$. Although DO concentrations were under saturation these were not low enough to warrant anaerobic conditions in the Lagoon, thus probably methane decreases. The more oxic the wetland is, the more CH_4 oxidation outweighs methanogenesis (Mitsch and Gosselink , 2007).

Organic matter concentrations (COD) influenced methane emissions at P1 and P2. In these points, there was entrance of organic matter from wastewater treatment and Cauca River. In addition, sediments from P1 and P2 showed higher COD concentrations than P3 and P4. This availability of substrate in Sonso Lagoon might explain in part higher methane emissions from P1 and P2. Methane is produced by re-mineralization of carbon accumulated in the sediments under anaerobic conditions, which is after emitted to the atmosphere. This indicates that CH_4 emissions from Sonso Lagoon may be affected by the input of anthropogenic carbon.

The presence of PO_4^{-3} and NH_4^+ was positively correlated to CH_4 fluxes while apparently higher NO_3^- coincided with lower CH_4 fluxes. Higher PO_4^{-3} concentrations coincided with higher CH_4 emissions measured in the Sonso Lagoon, as was reported for lakes and ditches from The Netherlands and Sweden (Johansson et al., 2004; Schrier-Uijl et al., 2011). An increment of PO_4^{-3} in lakes under eutrophication stimulates the transformation of organic matter to methane (Adhya et al., 1998; Huttunen et al., 2003; Conrad and Klose, 2005; Sun et al., 2013). In the same way, the results indicate that an increase in NH_4^+ increased CH_4 fluxes from the Sonso Lagoon. This increase may be explained because elevated NH_4^+ concentration inhibits CH_4 oxidation (Biswas et al., 2007; Borrel et al., 2011). This inhibition is attributed to competition between NH_4 and CH_4 for binding sites on methane monooxygenase, because of their similar chemical structure (Bédard and Knowles, 1989). The negative correlation between NO_3^- and methane production found in this study has also been reported in previous studies (Johansson et al., 2004; Søvik and Kløve, 2007).

6.4.2 Nitrous oxide dynamics

N_2O fluxes in the Sonso Lagoon were mostly higher than in boreal ponds and freshwater wetlands (Huttunen et al., 2002; Tremblay et al., 2004; Song et al., 2006; Yang et al., 2013) and matched with constructed wetlands that receive sewage (Johansson et al., 2003; Liikanen et al., 2006; Søvik and Kløve, 2007; Ström et al., 2007). This suggests that in the Sonso Lagoon, the unintentional runoff of N from fertilized fields, and human and livestock waste, has affected the potential of this aquatic system to retain and denitrify reactive nitrogen, leading to an increase in N_2O emission. Additionally, the results found in the current study are in agreement with previous studies in Finland and Switzerland that demonstrated an increasing trend of N_2O emissions when lakes change from oligotrophy to eutrophic conditions (Huttunen et al., 2003).

The atmospheric flux of N_2O in the wetland was positively correlated to TN and NO_3^-. Nitrogen had significant influences on N_2O production in an eutrophic lake (Wang et al., 2007). N_2O is an intermediate of both nitrification and denitrification and its atmospheric release depends on the availability of N (NO_3^-, NO_2^-, or NH_4^+) and oxygen (Groffman, 1991; Morris, 1991). The presence of NO_3, NO_2, or NH_4^+ is related to wastewater discharges, sediment exchange from the Cauca River, and agricultural run-off. Nitrifying bacteria may produce NO and N_2O either as a side-product in the catabolic pathway (oxidizing ammonia to nitrite), or, alternatively, denitrifying bacteria may produce NO or N_2O converting nitrite with ammonia, hydrogen or pyruvate as an electron donor (Colliver and Stephenson 2000; Wrage *et al.* 2001; Law *et al.* 2012). Nevertheless, it is difficult to conclude what the main processes are for N_2O production in the present study as nitrification and denitrification rates were not estimated.

Mean N_2O fluxes from the planktonic zone P3 were higher than P1, P2, and P4. The highest N_2O emission value at P3 was observed during July, the warmest month of the season. This coincided with the period of highest primary productivity (350 mg $C.m^{-3}.h^{-1}$) and highest Chl-a values of 84.7 $mg.m^{-3}$. In addition, at P3 a high abundance of *Trachelomonas, Cryptomonas sp and Tribonema minus* algae was observed which is typical of eutrophic aquatic ecosystems. These observations suggest that algae play an essential role in N_2O fluxes. According to Vörös *et al.*, (2003) the decomposition of planktonic algae likely released more soluble inorganic nutrients, thus stimulating N_2O production. The DO concentrations in P3 during daylight reached values of 5.4 mg. l^{-1}, which probably decreased the denitrification rate and blocked N_2O reduction to N_2. Further study is required to understand the actual mechanism of N_2O production related to algae presence, including comparison of day-time (photo-period) and night-time (anoxic) conditions.

The negative fluxes observed in the Sonso Lagoon were also reported for tidal mangrove wetlands (Wang *et al.*, 2016), eutrophic lakes (Wang *et al.*, 2007), coast salt marshes (Yuan *et al.*, 2015) and free water surface constructed wetlands (Johansson *et al.*, 2003; Liikanen *et al.*, 2006; Ström *et al.*,2007). In the current study, it was found that N_2O was taken up in the zone dominated by floating plants and the transport of this gas into the atmosphere was limited. This result is in line with observations in a constructed wetland treating wastewater (Johansson *et al.* 2003). In the zones with vegetation N_2O is both emitted and sequestered depending on the type of vegetation. Besides, N_2O consumption has also been related to NO_3^- concentration below 0.5 $mg.l^{-1}$ (Johansson *et al.*, 2003). The shortage of electron acceptors may have stimulated N_2O consumption by denitrifying bacteria. In the Sonso Lagoon, measurements indicated that NO_3^- was around 0.10 $mg.l^{-1}$, which may explain why the emission of N_2O into the atmosphere was low. However, it is necessary to elucidate the exact mechanism that affects N_2O consumption and its relation to other environmental factors and vegetation.

6.5 CONCLUSIONS

- The current study focused on greenhouse gas emissions from a eutrophic tropical freshwater wetland located in Colombia. The measurements over a period of one year showed an influence of anthropogenic activities such as agricultural run-off and wastewater discharges on the natural wetland switching its ability to sequester C and N to becoming sources of greenhouse gases. Thus, anthropogenic activities enhance the contribution of freshwater wetlands to global warming.

- In addition, the findings of this research suggest a role of wetland vegetation i.e. *Eichhornia crassipe* on GHG emissions. However, there is a need for further studies to elucidate the exact mechanisms of GHG emissions related to vegetation presence in tropical wetlands.

6.6 ACKNOWLEDGEMENTS

The author acknowledges the financial assistance of the EU funded SWITCH project for his PhD fellowship. The author is also indebted to the Universidad del Valle for its support. Recognition also goes to the students in sanitary and environmental engineering Ana Lasso, Teresita Canchala, Mayulli Gallardo and Ronny Nuñez for their assistance.

6.7 REFERENCES

Adhya, T., Pattnaik, P., Satpathy, S., Kumaraswamy, S. and Sethunathan, N. (1998). Influence of phosphorus application on methane emission and production in flooded paddy soils. Soil Biology and Biochemistry 30(2), 177-181.

Ahlgren, J., Reitzel, K., De Brabandere, H., Gogoll, A. and Rydin, E. (2011). Release of organic P forms from lake sediments. Water research 45(2), 565-572.

Anderson, B., Bartlett, K., Frolking, S., Hayhoe, K., Jenkins, J. and Salas, W. (2010). Methane and nitrous oxide emissions from natural sources. United States Environmental Protection Agency, Office of Atmospheric Programs, Washington DC.

APHA (2005) APHA,AWWA, WEF, Standard methods for the examination of water and wastewater, American Public Health Association, American Water Works Association and Water Environment Federation, 21 st ed. Washington DC.

Attermeyer, K., Flury, S., Jayakumar, R., Fiener, P., Steger, K., Arya, V., Wilken, F., van Geldern, R. and Premke, K. (2016). Invasive floating macrophytes reduce greenhouse gas emissions from a small tropical lake. Scientific reports 6, 20424 DOI: 20410.21038/srep20424.

Beaulieu, J.J., Tank, J.L., Hamilton, S.K., Wollheim, W.M., Hall, R.O., Mulholland, P.J., Peterson, B.J., Ashkenas, L.R., Cooper, L.W. and Dahm, C.N. (2011). Nitrous oxide emission from denitrification in stream and river networks. Proceedings of the National Academy of Sciences 108(1), 214-219.

Bédard, C. and Knowles, R. (1989). Physiology, biochemistry, and specific inhibitors of CH_4, NH_4^+, and CO oxidation by methanotrophs and nitrifiers. Microbiological reviews 53(1), 68-84.

Bellido, J.L., Peltomaa, E. and Ojala, A. (2011). An urban boreal lake basin as a source of CO_2 and CH_4. Environmental Pollution 159(6), 1649-1659.

Bergström, I., Mäkelä, S., Kankaala, P. and Kortelainen, P. (2007). Methane efflux from littoral vegetation stands of southern boreal lakes: An upscaled regional estimate. Atmospheric environment 41(2), 339-351.

Biswas, H., Mukhopadhyay, S.K., Sen, S. and Jana, T.K. (2007). Spatial and temporal patterns of methane dynamics in the tropical mangrove dominated estuary, NE coast of Bay of Bengal, India. Journal of Marine Systems 68(1–2), 55-64.

Bolpagni, R., Pierobon, E., Longhi, D., Nizzoli, D., Bartoli, M., Tomaselli, M. and Viaroli, P. (2007). Diurnal exchanges of CO_2 and CH_4 across the water–atmosphere interface in a water chestnut meadow (*Trapa natans L.*). Aquatic Botany 87(1), 43-48.

Borrel, G., Jézéquel, D., Biderre-Petit, C., Morel-Desrosiers, N., Morel, J.-P., Peyret, P., Fonty, G. and Lehours, A.-C. (2011). Production and consumption of methane in freshwater lake ecosystems. Research in Microbiology 162(9), 832-847.

Boström, B., Andersen, J.M., Fleischer, S. and Jansson, M. (1988). Exchange of phosphorus across the sediment-water interface. Hydrobiologia 170(1), 229-244.

Bridgham, S.D., Cadillo-Quiroz, H., Keller, J.K. and Zhuang, Q. (2013). Methane emissions from wetlands: biogeochemical, microbial, and modeling perspectives from local to global scales. Global Change Biology 19(5), 1325-1346.

Brix, H., Sorrell, B.K. and Lorenzen, B. (2001). Are Phragmites-dominated wetlands a net source or net sink of greenhouse gases? Aquatic Botany 69(2-4), 313-324.

Brix, H., Sorrell, B.K. and Schierup, H.-H. (1996). Gas fluxes achieved by in situ convective flow in *Phragmites australis*. Aquatic Botany 54(2), 151-163.

Casper, P., Maberly, S.C., Hall, G.H. and Finlay, B.J. (2000). Fluxes of methane and carbon dioxide from a small productive lake to the atmosphere. Biogeochemistry 49(1), 1-19.

Chowdhury, T.R. and Dick, R.P. (2013). Ecology of aerobic methanotrophs in controlling methane fluxes from wetlands. Applied soil ecology 65, 8-22.

Christensen, P.B., Nielsen, L.P., Sørensen, J. and Revsbech, N.P. (1990). Denitrification in nitrate-rich streams: diurnal and seasonal variation related to benthic oxygen metabolism. Limnology and Oceanography 35(3), 640-651.

Colliver, B. and Stephenson, T. (2000). Production of nitrogen oxide and dinitrogen oxide by autotrophic nitrifiers. Biotechnology Advances 18(3), 219-232.

Conrad, R. and Klose, M. (2005). Effect of potassium phosphate fertilization on production and emission of methane and its ^{13}C-stable isotope composition in rice microcosms. Soil Biology and Biochemistry 37(11), 2099-2108.

Davidson, T.A., Audet, J., Svenning, J.C., Lauridsen, T.L., Søndergaard, M., Landkildehus, F., Larsen, S.E. and Jeppesen, E. (2015). Eutrophication effects on greenhouse gas fluxes from shallow-lake mesocosms override those of climate warming. Global Change Biology 21(12), 4449-4463.

Deborah, C. (1996). Water quality assessments-A guide to use of biota, sediments and water in environmental Monitoring. UNESCO/WHO/UNEP.

Duan, X., Wang, X., Mu, Y. and Ouyang, Z. (2005). Seasonal and diurnal variations in methane emissions from Wuliangsu Lake in arid regions of China. Atmospheric environment 39(25), 4479-4487.

Groffman, P.M. (1991). Ecology of nitrification and denitrification in soil evaluated at scales relevant to atmospheric chemistry. Rogers, J. and Whitman, W. (eds) Microbial production and consumption of greenhouse gases: methane, nitrogen oxides, and halomethanes, American Society of Biology, Washington DC, 201-217.

Harrison, J.A., Matson, P.A. and Fendorf, S.E. (2005). Effects of a diel oxygen cycle on nitrogen transformations and greenhouse gas emissions in a eutrophied subtropical stream. Aquatic Sciences 67(3), 308-315.

Hendzel, L., Matthews, C., Venkiteswaran, J., St. Louis, V., Burton, D., Joyce, E. and Bodaly, R. (2005). Nitrous oxide fluxes in three experimental boreal forest reservoirs. Environmental Science and Technology 39(12), 4353-4360.

Huttunen, J.T., Alm, J., Liikanen, A., Juutinen, S., Larmola, T., Hammar, T., Silvola, J. and Martikainen, P.J. (2003). Fluxes of methane, carbon dioxide and nitrous oxide in boreal lakes and potential anthropogenic effects on the aquatic greenhouse gas emissions. Chemosphere 52(3), 609-621.

Huttunen, J.T., Hammar, T., Alm, J., Silvola, J. and Martikainen, P.J. (2001). Greenhouse gases in non-oxygenated and artificially oxygenated eutrophied lakes during winter stratification. Journal of Environmental Quality 30(2), 387-394.

Huttunen, J.T., Väisänen, T.S., Heikkinen, M., Hellsten, S., Nykänen, H., Nenonen, O. and Martikainen, P.J. (2002). Exchange of CO_2, CH_4 and N_2O between the atmosphere and two northern boreal ponds with catchments dominated by peatlands or forests. Plant and Soil 242(1), 137-146.

Johansson, A.E., Gustavsson, M., Oquist, M.G. and Svensson, B.H. (2004). Methane emissions from a constructed wetland treating wastewater--seasonal and spatial distribution and dependence on edaphic factors. Water research 38(18), 3960-3970.

Johansson, A.E., Klemedtsson, A.K., Klemedtsson, L. and Svensson, B.H. (2003). Nitrous oxide exchanges with the atmosphere of a constructed wetland treating wastewater. Parameters and implications for emission factors. Tellus B 55(3), 737-750.

Laanbroek, H.J. (2010). Methane emission from natural wetlands: interplay between emergent macrophytes and soil microbial processes. A mini-review. Annals of Botany 105(1), 141-153.

Law, Y., Ye, L., Pan, Y. and Yuan, Z. (2012). Nitrous oxide emissions from wastewater treatment processes. Philosophical Transactions of the Royal Society B: Biological Sciences 367(1593), 1265-1277.

Liikanen, A., Huttunen, J.T., Karjalainen, S.M., Heikkinen, K., Väisänen, T.S., Nykänen, H. and Martikainen, P.J. (2006). Temporal and seasonal changes in greenhouse gas emissions from a constructed wetland purifying peat mining runoff waters. Ecological Engineering 26(3), 241-251.

Liikanen, A., Ratilainen, E., Saarnio, S., Alm, J., Martikainen, P.J. and Silvola, J. (2003). Greenhouse gas dynamics in boreal, littoral sediments under raised CO2 and nitrogen supply. Freshwater Biology 48(3), 500-511.

Minkkinen, K. and Laine, J. (2006). Vegetation heterogeneity and ditches create spatial variability in methane fluxes from peatlands drained for forestry. Plant and Soil 285(1-2), 289-304.

Mitsch, W., Bernal, B., Nahlik, A., Mander, Ü., Zhang, L., Anderson, C., Jørgensen, S. and Brix, H. (2013). Wetlands, carbon, and climate change. Landscape Ecology 28(4), 583-597.

Mitsch, W.J. and Gosselink, J.G. (2007) Wetlands. Hoboken, NJ: John Wiley & Sons, Inc.

Morris, J.T. (1991). Effects of nitrogen loading on wetland ecosystems with particular reference to atmospheric deposition. Annual Review of Ecology and Systematics 22(1), 257-279.

Nicholls, R.J., Wong, P.P., Burkett, V., Codignotto, J., Hay, J., McLean, R., Ragoonaden, S., Woodroffe, C.D., Abuodha, P. and Arblaster, J. (2007). Climate change 2007: impacts, adaptation and vulnerability. Contribution of working group II to the fourth assesment report of the IPCC, Intergovermental Panel on Climate Change, Cambridge, UK.

Peixoto, R., Marotta, H., Bastviken, D. and Enrich-Prast, A. (2016). Floating Aquatic Macrophytes Can Substantially Offset Open Water CO2 Emissions from Tropical Floodplain Lake Ecosystems. Ecosystems, 1-13.

Raymond, P.A., Hartmann, J., Lauerwald, R., Sobek, S., McDonald, C., Hoover, M., Butman, D., Striegl, R., Mayorga, E. and Humborg, C. (2013). Global carbon dioxide emissions from inland waters. Nature 503(7476), 355-359.

Sánchez-Carrillo, S., Angeler, D., Álvarez-Cobelas, M. and Sánchez-Andrés, R. (2011). Eutrophication: causes, consequences and control, pp. 195-210, Springer.

Schedlbauer, J.L., Munyon, J.W., Oberbauer, S.F., Gaiser, E.E. and Starr, G. (2012). Controls on ecosystem carbon dioxide exchange in short-and long-hydroperiod Florida Everglades freshwater marshes. Wetlands 32(5), 801-812.

Schrier-Uijl, A., Veraart, A., Leffelaar, P., Berendse, F. and Veenendaal, E. (2011). Release of CO_2 and CH_4 from lakes and drainage ditches in temperate wetlands. Biogeochemistry 102(1), 265-279.

Serediak, N.A. (2014). A critical examination of chemical extremes in freshwater systems, Lakehead University.

Silva, J.P., Lasso, A., Lubberding, H.J., Peña, M.R. and Gijzen, H.J. (2015). Biases in greenhouse gases static chambers measurements in stabilization ponds: Comparison of flux estimation using linear and non-linear models. Atmospheric environment 109, 130-138.

Silva, J.P., Ruiz, J.L., Peña, M.R., Lubberding, H. and Gijzen, H. (2012). Influence of photoperiod on carbon dioxide and methane emissions from two pilot-scale stabilization ponds. Water Science & Technology 66(9).

Smith, T.M. and Smith, R.L. (2007) Elementi di ecologia, Pearson.

Søndergaard, M. (2007). Nutrient dynamics in lakes-with emphasis on phosphorus, sediment and lake restoration, Aarhus Universitet, Danmarks Miljøundersøgelser, Afdeling for Ferskvandsøkologi.

Song, C., Yang, G., Liu, D. and Mao, R. (2012). Phosphorus availability as a primary constraint on methane emission from a freshwater wetland. Atmospheric environment 59, 202-206.

Song, C., Zhang, L., Wang, Y. and Zhao, Z. (2006). Annual dynamics of CO_2, CH_4, N_2O emissions from freshwater marshes and affected by nitrogen fertilization. Huan jing ke xue 27(12), 2369-2375.

Sorrell, B. and Brix, H. (2003). Effects of water vapour pressure deficit and stomatal conductance on photosynthesis, internal pressurization and convective flow in three emergent wetland plants. Plant and Soil 253(1), 71-79.

Søvik, A. and Kløve, B. (2007). Emission of N_2O and CH_4 from a constructed wetland in southeastern Norway. Science of The Total Environment 380(1), 28-37.

Ström, L., Lamppa, A. and Christensen, T.R. (2007). Greenhouse gas emissions from a constructed wetland in southern Sweden. Wetlands ecology and management 15(1), 43-50.

Sun, Q.-Q., Shi, K., Damerell, P., Whitham, C., Yu, G.-H. and Zou, C.-L. (2013). Carbon dioxide and methane fluxes: Seasonal dynamics from inland riparian ecosystems, northeast China. Science of The Total Environment 465, 48-55.

Teiter, S. and Mander, U. (2005). Emission of N_2O, N_2, CH_4, and CO_2 from constructed wetlands for wastewater treatment and from riparian buffer zones. Ecological Engineering 25(5), 528-541.

Tremblay, A., Lambert, M. and Gagnon, L. (2004). Do hydroelectric reservoirs emit greenhouse gases? Environmental Management 33, 509-517.

Van der Steen, N.P., Ferrer, A.V.M., Samarasinghe, K.G. and Gijzen, H.J. (2003). Quantification and comparison of methane emissions from algae and duckweed based wastewater treatment ponds. Universidad del Valle; CINARA; International Water Association. Memorias del evento: Agua 2003. Cartagena de Indias, IWA, 2003, p. 1-7.

Ventura, R.E. (2014). Wetlands and Greenhouse Gas Fluxes: Causes and Effects of Climate Change–A Meta-Analysis, Pomona College.

Vörös, L., Katalin, V., Koncz, E. and Kovács, A. (2003). Phytoplankton and bacterioplankton production in a reed-covered water body. Aquatic Botany 77(2), 99-110.

Wang, H., Liao, G., D'Souza, M., Yu, X., Yang, J., Yang, X. and Zheng, T. (2016). Temporal and spatial variations of greenhouse gas fluxes from a tidal mangrove wetland in Southeast China. Environmental Science and Pollution Research 23(2), 1873-1885.

Wang, H., Yang, L., Wang, W., Lu, J. and Yin, C. (2007). Nitrous oxide (N_2O) fluxes and their relationships with water-sediment characteristics in a hyper-eutrophic shallow lake, China. Journal of geophysical research 112(G1), G01005.

Wang, Y., Yang, H., Ye, C., Chen, X., Xie, B., Huang, C., Zhang, J. and Xu, M. (2013). Effects of plant species on soil microbial processes and CH 4 emission from constructed wetlands. Environmental Pollution 174, 273-278.

Whalen, S. (2005). Biogeochemistry of methane exchange between natural wetlands and the atmosphere. Environmental Engineering Science 22(1), 73-94.

Wrage, N., Velthof, G., Van Beusichem, M. and Oenema, O. (2001). Role of nitrifier denitrification in the production of nitrous oxide. Soil Biology and Biochemistry 33(12), 1723-1732.

Wright, A.L., Ramesh Reddy, K.R. and Newman, S. (2009). Microbial indicators of eutrophication in Everglades wetlands. Soil Science Society of America Journal 73(5), 1597-1603.

Wright, A.L., Reddy, K.R. and Newman, S. (2008). Biogeochemical response of the Everglades landscape to eutrophication. Global J Environ Res 2, 102-109.

Xing, Y., Xie, P., Yang, H., Ni, L., Wang, Y. and Rong, K. (2005). Methane and carbon dioxide fluxes from a shallow hypereutrophic subtropical Lake in China. Atmospheric environment 39(30), 5532-5540.

Xing, Y., Xie, P., Yang, H., Wu, A. and Ni, L. (2006). The change of gaseous carbon fluxes following the switch of dominant producers from macrophytes to algae in a shallow subtropical lake of China. Atmospheric environment 40(40), 8034-8043.

Yang, J., Liu, J., Hu, X., Li, X., Wang, Y. and Li, H. (2013). Effect of water table level on CO_2, CH_4 and N_2O emissions in a freshwater marsh of Northeast China. Soil Biology and Biochemistry 61, 52-60.

Yang, L., Lu, F., Zhou, X., Wang, X., Duan, X. and Sun, B. (2014). Progress in the studies on the greenhouse gas emissions from reservoirs. Acta Ecologica Sinica 34(4), 204-212.

Yuan, J., Ding, W., Liu, D., Kang, H., Freeman, C., Xiang, J. and Lin, Y. (2015). Exotic Spartina alterniflora invasion alters ecosystem–atmosphere exchange of CH_4 and N_2O and carbon sequestration in a coastal salt marsh in China. Global Change Biology 21(4), 1567-1580.

Zedler, J.B. and Kercher, S. (2005). Wetland resources: status, trends, ecosystem services, and restorability. Annu. Rev. Environ. Resour. 30, 39-74.

Chapter 7

General discussion, outlook and conclusions

7.1 INTRODUCTION

Ecotechnologies for wastewater treatment (EWWT) have been used as a cost-effective alternative to conventional wastewater treatment methods for improving the removal of organic carbon, nutrients and pathogenic microorganisms from wastewater. However, due to biochemical transformations of organic matter an nutrients EWWT are net sources of CO_2, CH_4 and N_2O GHs which may be transferred into the atmosphere contributing to global warming (Daelman et al., 2013).

Greenhouse gases such as CO_2, CH_4 and N_2O have been observed in anaerobic ponds (Toprak, 1995; Picot et al., 2003; Wang et al., 2011; Konaté et al., 2013; Paredes et al., 2015), facultative ponds (Stadmark and Leonardson, 2005; Detweiler et al., 2014; Hernandez-Paniagua et al., 2014; Glaz et al., 2016), and constructed wetlands (Tanner et al., 1997; Fey et al., 1999; Johansson et al., 2003; Johansson et al., 2004; Mander et al., 2005; Teiter and Mander, 2005; Liikanen et al., 2006; Sovik et al., 2006; Gui et al., 2007; Søvik and Kløve, 2007; De Klein and Van der Werf, 2014; Wu et al., 2016). Thus, there is a concern that implementing EWWT can turn into an atmospheric pollution problem and climate change impact.

Because there is a global concern about the possible effects of human activities on global warming, the identification of GHGs from EWWT remains to be answered. Although it has been estimated that the wastewater sector accounts for 3-4 % of greenhouse gas global emissions (Bogner et al., 2007; Ciais et al., 2014; Saunois et al., 2016), there are knowledge gaps mainly regarding data on GHG emissions from tropical regions. There is also a lack of understanding of CO_2, CH_4, and N_2O dynamics under tropical conditions. The tropical conditions are characterized by high temperatures, long and stable photoperiods, photosynthetic activity, high levels of bacterial and algal activity, and dynamics in dissolved oxygen and pH patterns which may influence GHG dynamics differently compared to temperate conditions.
In this final chapter, the main results obtained are summarized and discussed to provide an overview of the research performed in this dissertation.

7.2 FLUX ESTIMATION OF GREENHOUSE GASES FROM STABILIZATION PONDS USING THE STATIC CHAMBER TECHNIQUE: COMPARISON OF LINEAR AND NON-LINEAR MODELS

The first part of the research carried out in this thesis (Chapter 2) focused on the adaptation and assessment of a measurement technique to estimate the GHG flux produced in EWWT. The static chamber technique was chosen because it has been the most useful and reported technique in measuring GHG from wastewater plants, aquatic ecosystems and soils. The results obtained in Chapter 2 suggest that closed static chambers are a good analytical methodology to estimate GHG emissions from wastewater stabilization ponds. However, there are strong recommendations for analysing data and operational conditions for the chamber during the measurement of GHG.

The literature survey revealed a gap in the flux calculation using the static chamber technique. In most of the studies reported, this calculation was based on the assumption of a linear increase in the concentration of the different GHGs in the chamber headspace. The linear approach was not acceptable because of the non-steady-state conditions of closed static chambers – and most likely also of the different processes - occurring in wastewater treatment (Pedersen *et al.*, 2010). The result of this inaccuracy in the basic linear assumption could lead to an underestimation of GHG fluxes from wastewater treatment systems.

The hypothesis developed in Chapter 2 shows that the measurements using static chambers to estimate GHG fluxes from EWWT are not completely linear. Six different kinds of behaviour were observed for GHG concentration over time in the headspace of chambers during the measurement time (see Fig. 2.2). These behaviours are influenced by potential error sources related to water surface disturbances, temperature, concentration, and pressure gradients within the chamber, moisture saturation and lack of mixing in the headspace.

The data analysis indicated that the linear model might be applied to estimate fluxes only in some instances (60% data). By contrast, approximately 40% of the data were best fitted to a non-linear regression model (quadratic or exponential). Thus, because almost all the trends are not linear, three different approaches should be used and compared in the flux calculations:

$$C = a_0 + a_1 t \quad \text{Linear}$$
$$C = a_0 + bt + ct^2 \quad \text{Quadratic}$$

$$C = C_{max} - (C_{max} - C_0) . \exp^{(-k t)} \quad \text{Exponential}$$

Further, the flux comparisons obtained by applying these regression models should be carried out based on the statistical criterion R^2_{adj}. The literature review evidenced that the coefficient R^2 has been the only criterion used to decide the data concentration goodness of fit when measured within the headspace of the chamber. From the current research, it is clear that when $R^2_{adj-non-lin} > R^2_{adj-lin}$, the application of linear regression models is not appropriate and may lead to an underestimation of GHG fluxes of between 10 and 50%. Therefore, the incorrect use of the usual R^2 parameter and only the linear regression model to estimate the fluxes may lead to severe underestimation of the real contribution of GHG emissions from wastewater.

7.3 EFFECT OF COD LOADING RATES AND NON-UNIFORM SLUDGE ACCUMULATION ON GREENHOUSE GAS EMISSIONS FROM ANAEROBIC PONDS

In anaerobic ponds (APs), organic matter removal is accomplished by a microbiological process leading to biogas production. This biogas is mainly composed of GHGs such as CH_4 and CO_2. On the one hand, when biogas is collected, it can be used to generate electricity and capture carbon. On the other hand, when this biogas is released from an anaerobic pond surface this wastewater treatment system contributes to global warming. Methane is the second most important anthropogenic greenhouse gas after CO_2, contributing approximately 30% to the total net anthropogenic radiative forcing of 1.6 $W.m^{-2}$ (Ciais et al., 2014; Saunois et al., 2016). Thus, anaerobic ponds could reduce the impact on one environmental factor i.e. water pollution while at the same time creating additional impacts in another area i.e. air pollution.

Undertaking a full-scale study on an anaerobic pond treating domestic wastewater was an attempt to provide performance data on GHG fluxes such as CO_2 and CH_4, and N_2O under tropical conditions (Chapter 4). The figures suggest that the AP studied emitted a substantial amount of CH_4 and CO_2 compared to those reported under Mediterranean and subtropical climatic conditions (Toprak, 1995; Picot et al., 2003; Hernandez-Paniagua et al., 2014), but lower than observed under Sudano-Sahelian climate (Konaté et al., 2013). From these results, it is clear that the highest temperature leads to the highest GHG emissions.

This study also corroborated that the organic loading rate (OLR) and COD influenced the production of GHGs in anaerobic ponds (**Table 3.2; Chapter 3**). It is expected that COD removal efficiency will increase with increasing organic loading rate and therefore more biogas production is observed. The production of biogas in Ginebra´s AP directly correlated to the COD removed, explaining the 64% of emissions produced ($p<0.05$). This was also reported in other APs that were inefficiently producing biogas during the holiday and rainy season where the pond operated with a low organic loading rate (Toprak, 1995; Konaté et al., 2013) Wastewater facilities treating effluents from the palm oil mill, livestock and tapioca industries showed a higher GHG production than was observed in the current study (Chapter 4). This is because anaerobic ponds that are designed to treat municipal sewage are projected to operate with a lower organic load than those anaerobic ponds treating industrial effluent.

Because the efficiency of COD removal in the AP was relatively low, the biogas production observed was also lower than theoretically expected. The average amount of CH_4 produced in the AP under tropical conditions (Ginebra) was 0.24±0.09 m^3. $kgCOD^{-1}_{removed}$ whereas CO_2 was approximately 0.18±0.07 m^3 $CO_2.kgCOD^{-1}_{removed}$. Based on the stoichiometry suggested by Metcalf and Eddy (2003) and under the temperature and atmospheric pressure conditions at Ginebra (T=26°C and b.p =0.89 atm.), these amounts were lower than theoretically expected of 0.43 m^3 $CH_4.kgCOD^{-1}_{removed}$ and 0.35 m^3 $CO_2 .kgCOD^{-1}_{removed}$.

The mass balance over the entire AP determined that approximately 37% of the influent COD that was converted into methane transferred into the atmosphere. In addition, 36% and 21% of COD was in both the effluent and settled solids, respectively. Finally, it was estimated that approximately 2.5% of the COD was CH_4 dissolved in the effluent of the AP whereas approximately 3.5% of the COD was in the volatile solids produced. From these results, the biogas recovery from the municipal wastewater through anaerobic processes might not be economically practical or environmentally friendly due to the fact that a substantial amount of methane is dissolved into the treated effluent. The dissolved methane would lead to reduced energy efficiency of the anaerobic process for municipal wastewater treatment and increase the risk of its release into the environment in the subsequent treatments i.e. secondary facultative ponds (SFPs). It is well known that biogas escaping into the atmosphere contributes to global warming (the greenhouse effect) and therefore, it is strongly suggested that APs should be designed to capture methane and to avoid GHG emissions.

Chapter 3 also revealed that the biogas production rate in an AP was limited by its excessive sludge accumulation. The organic material removal mechanism in the AP occurs because the liquid flows through the pond and the settleable influent material accumulates at the bottom, where the biodegradable fraction is digested by the anaerobic sludge mass. The sludge excess is produced by an insufficient operation and maintenance programme. In Ginebra's AP the last de-sludging was carried out five years before this study and the average sludge accumulation was 0.06 m^3 per person per year. In addition, the sludge was uneven distributed occurring mainly in the last third of the pond. As a consequence, the CH_4 and CO_2 emissions were lower in the AP outlet than the other AP zones i.e. the entrance or central zone. This could be because the outlet zone was dominated by sludge accumulation of high-density (mineralized), which is not easily biodegradable in a large quantity and therefore the biogas production decreased. Further, the sludge accumulation affected the hydraulic behaviour of the AP, the retention time is probably decreased and of course the effective reaction volume reduces, leading to a low removal rate of COD. These results address the importance of implementing best practices for O&M mainly when APs are projected to recover energy from biogas.

Considering that N_2O is a potent greenhouse gas that contributes significantly to global warming, the emissions of this gas were measured in an AP (Chapter 3). The median emission observed of N_2O was 6.8 ± 3.6 ml.m^{-2}. d^{-1} (12.2 ± 6.4 mg.m^{-2}. d^{-1}). Because N_2O has not been extensively reported as being emitted from anaerobic ponds, it was not possible to compare the results found in this research with data from other similar wastewater treatment systems. However, these findings suggest that N_2O production should not be excluded when determining greenhouse gas emissions in anaerobic ponds.

7.4 GHG DYNAMICS FROM SECONDARY FACULTATIVE STABILIZATION PONDS

The influence of photoperiod and wastewater characteristics on greenhouse gas emissions such as CH_4, CO_2 and N_2O was studied in secondary facultative lagoons (SFPs). On the one hand, GHG emissions from two types of small pilot-scale SFPs, algae facultative pond (AFP) and duckweed-based pond (DBP), were determined under daytime and night-time conditions (Chapter 4). On the other hand, in a second study (Chapter 5) the daily variation in GHG emissions was determined through different hours of the day in a full-scale AFP.

7.4.1 General overview of daily variability of GHGs emitted from secondary facultative stabilization ponds

The comparison of CO_2, CH_4, and N_2O fluxes measured in the SFPs studied (**Table 7.1**) shows a large daily variability (Chapter 4 and Chapter 5). The mean daytime and night-time fluxes indicated that the SFPs were sources of CO_2, CH_4, and N_2O (**Table 7.1**). Chapter 4 and Chapter 5 reveal that there was some consumption of CO_2, CH_4, and N_2O in the SFPs. The CO_2 and CH_4 sequestration were observed during the daytime whereas N_2O capture happened in both the daytime and night-time. Despite this consumption, it was not enough to offset the GHG emissions from the SFPs (**Table 7.1**).

In general, during the daytime the CO_2 fluxes (**Table 7.1**) were almost all negative whereas during the night-time CO_2 fluxes were positive (Chapter 4 and Chapter 5). CO_2 uptake by algal photosynthesis was only measured and reported in Australian and Canada SFPs (Hernandez-Paniagua *et al.*, 2014; Glaz *et al.*, 2016). Compared to temperate wastewater treatment systems, CO_2 fluxes were 8 and 20-fold lower than those in constructed wetlands (Liikanen *et al.*, 2006; Sovik *et al.*, 2006; Ström *et al.*, 2007). Likewise, the CO_2 consumption and emissions in the present study were 2 and 6-fold higher than those reported in shallow and deep lakes (Tremblay *et al.*, 2004; Xing *et al.*, 2005). The difference between tropical and temperate conditions is likely related to environmental condition – temperature being an obvious factor.

Regarding methane fluxes, these were mainly positive, indicating that SFPs can be sources of this gas. However, it should be noted that during the daytime, consumption or capture of this gas was observed in both AFP and DBP pilot-scale (Chapter 4) whereas in full-scale AFP all the CH_4 fluxes were always positive (Chapter 5). When compared with other studies, the CH_4 mean fluxes from pilot-scale AFP were 14 times higher than from the Australian and Canadian SFPs (Glaz *et al.*, 2016), similar to values reported for HSSF and VSSF constructed wetlands operating in Europe (Johansson *et al.*, 2004; Sovik *et al.*, 2006; Ström *et al.*, 2007). Further, CH_4 from full-scale AFP were twofold higher than those of temperate ponds(Johansson *et al.*, 2004), ten-times higher than subtropical ponds (Hernandez-Paniagua *et al.*, 2014) and nearly a hundredfold higher than boreal ponds (Glaz *et al.*, 2016). Further, as in the current study, some CH_4 sink behaviour was observed in an AFP system treating municipal wastewater (Johansson *et al.*, 2004; Detweiler *et al.*, 2014). On the other hand, it has been reported that the presence of a duckweed cover reduces GHG emissions i.e. CH_4 in DBPs (Van der Steen *et al.*, 2003).

Table 7.1 CO_2, CH_4, and N_2O emissions from secondary stabilization ponds measured in different photoperiods (daytime and night-time). The number in parenthesis is the median.

Period	SFP type	CO_2 mg.m^{2}d^{-1}	CH_4 mg.m^{2}d^{-1}	N_2O mg.m^{2}d^{-1}
Daytime	AFP pilot-scale	-2,963 - 6,403 (-232)	-80 - 200 (9.9)	-4.1 - 63.5 (6.9)
	DBP pilot-scale	-5,039.2 - 4,439 (-1,654.5)	-40 - 250 (71.4)	1.6 - 42.3 (8.5)
	AFP full-scale	-3,100 - 300 (-743)	290 - 4,510 (2,466)	-6.3 to 3.6 (-0.95)
Night-time	AFP pilot-scale	1,061-7,655 (3,950)	9 - 388 (12.7)	-8.8 - 40.5 (5.5)
	DBP pilot-scale	3,101-9,898 (5,116)	8.4 - 880 (195.2)	-16 -23.8 (2)
	AFP full-scale	1,040 - 4,730 (2,497)	80 - 3,850 (2,254)	-2.4 - 15 (3.8)

The median values of N_2O suggest that this gas was emitted from AFP and DBP pilot-scale during both the daytime and night-time. The mean values in pilot-scale and full-scale AFP investigated were higher than those found in SFPs in Quebec (Glaz *et al.*, 2016), subtropical SFPs (Hernandez-Paniagua *et al.*, 2014), a tropical urban pond (Singh *et al.*, 2005), constructed wetlands (Fey *et al.*, 1999; Johansson *et al.*, 2003; Teiter and Mander, 2005; Liikanen *et al.*, 2006; Sovik *et al.*, 2006; Ström *et al.*, 2007) and natural systems (Huttunen *et al.*, 2002; Huttunen *et al.*, 2003; Hopfensperger *et al.*, 2009). Moreover, in full-scale AFP studied, negative N_2O flux was consistent with the data measured in Swedish ponds (Johansson *et al.*, 2003; Ström *et al.*, 2007).

The results found in the current research (Chapter 4 and Chapter 5) and reported in the literature, open a discussion about bias if whole SFPs CO_2, CH_4, and N_2O flux estimates are based on daytime measurements only. This is because of the difference in GHGs uptake and release between the daytime and night-time in SFPs. The estimation of fluxes from SFPs may be underestimated and the actual contribution of SFPs to the global GHG budget may be more significant. Thus, the measurements of GHGs in SFPs must involve the influence of the photoperiod and wastewater characteristics.

7.4.2 Influence of photoperiod and wastewater characteristics on carbon dioxide, methane and nitrous oxide emissions from SFPs.

In Chapter 4 and Chapter 5 a statistical analysis revealed (i) the complex dynamic in SFPs due to changes in the forcing functions such as pH, DO, and temperature that are influenced by light intensity (photoperiod), and that (ii) GHG emissions are influenced by wastewater characteristics.

Environmental drivers of CO_2 flux variability in SFPs

The regression model (Table 5.4 in Chapter 5) indicates that CO_2 fluxes correlated positively with COD and negatively with pH and DO. The model therefore clarifies the 74% variation in CO_2 distribution. An analysis of the individual contribution for each environmental parameter indicated that pH and DO fluctuations account for 37.5% and 31%, respectively, of the CO_2 variations. Nonetheless, COD could only account for 5.5%.

There was a negative correlation between pH and CO_2. During the daytime, the pilot-scale AFP and full-scale AFP showed pH values between 8.5 and 9 units (Fig. 5.3 in Chapter 5). Algae require large quantities of dissolved carbon dioxide during photosynthesis, causing a depletion in CO_2 and leading to a shift in the carbonate-bicarbonate (CO_3^{-2} and HCO_3^-) equilibrium, resulting in an increase in pH due to the formation of hydroxyl (OH^-) ions (Kayombo et al., 2002). Thus, the formation of carbonic acid and bicarbonate leads to an under saturation of dissolved CO_2 in the water column and enhances the net transfer of CO_2 from the atmosphere into the water. The result of this is an increase in sequestration capacity by algae during the daytime. In contrast, at night-time algae and duckweed switch to respiration and hence produce CO_2 in addition to the CO_2 rendered by heterotrophic bacteria sufficient to generate CO_2 super-saturation in the water column, increasing the emissions of this gas into the atmosphere.

In pilot-scale DBPs (Chapter 4) there was higher CO_2 consumption than observed in both pilot-scale and full-scale AFPs. However, this consumption could not be correlated to pH because this environmental parameter was almost constant and close to neutrality (7.2±0.2). Duckweed achieves optimal growth and productivity in nutrient-rich waters and when it is exposed to an adequate light intensity. Under tropical conditions there is more light intensity in the red and blue spectrum that stimulates the photosynthetic pigments (chlorophylls and carotenoids) to trap more CO_2. Thus, there was probably more photosynthesis from duckweed than algae.

Environmental drivers of CH_4 variability in SFPs

Regarding CH_4, this gas in the AFP evaluated was positively correlated to TKN and ambient temperature (Table 5.4 of Chapter 5). However, these environmental parameters could only explain 52% of the CH_4 variations found in AFP. The statistical analysis revealed that TKN accounted for 38.7% variations in methane gas, while ambient temperature accounted for 13.3%. Regrettably, no plausible hypothesis for the influence of water temperature on CH_4 changes in AFP studied can be adduced.

The diurnal evolution of the methane flux indicates that emissions of this gas could be influenced by DO concentrations in the AFP and by duckweed cover in the DBP. Chapter 4 and Chapter 5 reveal that the AFP studied could achieve dissolved oxygen concentrations higher than 13.2 mg.L^{-1} during the daytime period and lower than 1.8 mg.L^{-1} at nigh-ttime. Probably the high concentration of DO observed in the AFP studied (Figure 5.4 of Chapter 5) led to a slightly methane oxidation in the water column while low DO (1.4 mg.L^{-1} O$_2$) relates to negative ORP values (-260mV) favoured methane production and a subsequent release into

the atmosphere. In contrast, the higher biomass density of floating duckweed in DBPs may function as a physical barrier for gas transfer, thus preventing a major release of CH$_4$ as suggested previously by Van der Steen *et al.* (2003). However, during the daytime and night-time periods DBP showed higher emissions of CH$_4$ than those observed in AFP (Fig. 4.2 and 4.4 in Chapter 4).

Because daytime and night-time CH$_4$ fluxes were always positive in the SFPs studied (Chapter 4 and Chapter 5), there may be other factors different to DO that limit the methane oxidation i.e. pond depth, organic load and dissolved CH$_4$. The full-scale AFP (Chapter 5) was deeper and a higher organic load was applied than the pilot-scale AFP and DBP, then there was a lower oxygen transfer into the water column prevailing anoxic conditions which favoured the methane production rather than methane oxidation. Furthermore, the highest emissions of CH$_4$ were observed in the entrance of the SFPs (Chapter 4), suggesting that some dissolved methane from the effluent of the anaerobic pond was released into the atmosphere. This situation describes the impacts of not having an adequate recovery of all the methane that is produced in anaerobic wastewater treatment systems.

Environmental drivers of N$_2$O variability in SFPs
The effects of the daily variation of pH and DO on N$_2$O emissions from SFPs have not been previously reported in SFPs under tropical conditions. Only a study on SFPs in Western Australia and Canada has reported GHG measurements for both daytime and night-time periods (Glaz *et al.*, 2016). In that study N$_2$O fluxes did not show any clear pattern and varied little over the 24-h cycle, except in only one SFP where N$_2$O peaked at 22:00 hours (Glaz *et al.*, 2016). Further, the fluxes measured were not correlated to pH or DO concentrations. In this context, it is evident that the results described in Chapter 4 and Chapter 5 help to better understand the dynamics of N$_2$O generation in SFPs.

According to the statistical test, the fluxes of N$_2$O in SFPs, were influenced by NO$_3^-$ (r^2=0.28), TKN (r^2=0.16), DO (r^2=0.14), COD (r^2=0.14), ambient temperature (r^2=0.12), water temperature (r^2=0.06), and NH$_4^+$ (r^2=0.005). These factors explained 85% of the variations of N$_2$O flux in SFPs (Table 5.5 in Chapter 5).

The elevation of the DO to super-saturation conditions (>13 mg.L^{-1}) during the largest solar radiation (9:00 to 15:00 hours) coincided with a decrease in N$_2$O production (Fig. 5.6 in Chapter 5). It is expected that under DO super-saturation conditions, some nitrification occurred in the SFPs studied. During nitrification, the DO concentration is a very important factor influencing

N$_2$O emissions, as higher DO levels lead to lower N$_2$O emissions (Johansson *et al.*, 2003; Law *et al.*, 2011; Guo *et al.*, 2018; Vieira *et al.*, 2018)

This decrease in N$_2$O emissions also coincided with when pH was between 8-9.0 units. Further, a decrease of DO to 1.2 ± 0.9 mgO$_2$.L^{-1} and pH of approximately 6.4 units decreased the N$_2$O production rate during night-time (00:00 to 06:00 hours). However, the N$_2$O production rate was the highest when DO was 4.5 mg.L^{-1} at pH 7.7 (21:00 hours). An increasing pH shifts the equilibrium to FA, which is the true substrate of AOB (Suzuki *et al.*, 1976; Law *et al.*, 2012), and is inhibitory to nitrite-oxidizing bacteria (NOB) (Vadivelu *et al.*, 2007). Conversely, a decreasing pH increases the FNA concentration, which inhibits both AOB and NOB (Vadivelu *et al.*, 2006).

The regression analysis revealed that NO$_3^-$ and NH$_4^+$ correlated positively and negatively, respectively, with the N$_2$O fluxes generated from SFPs (Table 5.5 in Chapter 5). When the concentration of NO$_3^-$ was increased the fluxes of N$_2$O into the atmosphere also were increased suggesting that denitrification was the mechanism dominant. By contrast, decreasing NH$_4^+$ during nitrification increased the production of N$_2$O. Although in Chapter 4 and Chapter 5 nitrification and denitrification processes were not assessed directly, these results indicate that N$_2$O production in SFPs may be mainly attributed to both nitrification and denitrification.

High oxygen concentrations that prevailed in the Ginebra AFP during the daytime facilitated ammonium removal via nitrification (Table 4.2 in Chapter 4). However, when anoxic conditions were likely dominant, N$_2$O could have been produced due to nitrifier- denitrification and incomplete oxidation of NH$_2$OH (Daelman *et al.*, 2015; Guo *et al.*, 2018). On the other hand, the availability of nitrate and low organic matter probably promotes high N$_2$O emission by incomplete denitrification. A number of factors relating to N$_2$O emissions in SFPs such as partial anaerobic conditions, availability of biodegradable organic substrate and moderated nitrate concentration, could be favourable. Thus, N$_2$O by heterotrophic denitrification could occur due to the fact that the AFP and DBP were operated at a low COD / N ratio and an average DO of 1.5 mg. L^{-1} (Itokawa *et al.*, 2001).

7.5 ANTHROPOGENIC INFLUENCE ON GREENHOUSE GAS FLUXES IN A EUTROPHIC TROPICAL FRESHWATER WETLAND

Chapter 6 assessed the dynamics of GHGs in a shallow eutrophic freshwater wetland under tropical conditions (Sonso Lagoon). This investigation was included in this research because tropical wetlands play an important role in the global carbon (C) cycle (Page *et al.*, 2011; Wright *et al.*, 2013; Sjögersten *et al.*, 2014). Further, considering that only about 9% of wastewater effluents are treated in Colombia, many freshwater wetlands (FWs) act as an indirect EWWT for domestic sewage treatment and therefore exhibit eutrophic conditions.

7.5.1 Context for the GHG emissions from a eutrophic tropical natural

Compared to the more intensively studied boreal and temperate FWs, tropical FWs have been poorly studied with regarding to GHG emissions. Thus, information on CO_2, CH_4 and N_2O

fluxes from these systems is lacking. In the literature, very few papers have examined N_2O in wetlands in comparison to studies about CO_2 and CH_4. In addition, in several studies it has been reported that the N_2O amount is negligible.

In the Sonso Lagoon, CH_4 fluxes ranged from -0.4 to 5.3 $g.m^{-2}.d^{-1}$. These figures were higher than observed in temperate forested peatland (Huttunen *et al.*, 2002; Tremblay *et al.*, 2004;

Xing *et al.*, 2005; Schrier-Uijl *et al.*, 2011) and tropical forested peatland (Melling *et al.*, 2005; Jauhiainen *et al.*, 2012). The results are comparable with those reported in forested peatland in Panama (Wright *et al.*, 2013) and flooded forest (Nahlik and Mitsch, 2011). Negative CH_4 fluxes from Sonso (i.e. uptake) were three-fold higher than those observed in other tropical wetlands (Wright *et al.*, 2013) and 100-fold higher than measured in tropical forested peatland (Jauhiainen *et al.*, 2005).

A high variability in CO_2 fluxes was observed within Sonso Lagoon. CO_2 fluxes varied between -2.4 and 11.5 $g.m^{-2}.d^{-1}$. The CO_2 exchange in wetlands has been reported in several studies but the magnitude of uptake of CO_2 is 2-25 times lower compared than the results obtained in Sonso (Huttunen *et al.*, 2002; Tremblay *et al.*, 2004; Xing *et al.*, 2005; Schrier-Uijl *et al.*, 2011). Surprisingly, in other tropical wetlands CO_2 consumption was not observed (Jauhiainen *et al.*, 2005; Melling *et al.*, 2005; Wright *et al.*, 2013). It may be that the opaque chamber used only measured emissions from the soil or without photosynthesis.

The estimated flux of N_2O ranged between -0.015 and 0.012 $g.m^{-2}.d^{-1}$. N_2O fluxes in Sonso Lagoon were mostly higher than in boreal ponds and freshwater wetlands (Huttunen *et al.*, 2002; Tremblay *et al.*, 2004; Song *et al.*, 2006; Yang *et al.*, 2013) and comparable to tropical mangroves (Krithika *et al.*, 2008), and subarctic peatland (Crill *et al.*, 1988). The N_2O emission also matched with constructed wetlands that receive sewage (Johansson *et al.*, 2003; Liikanen *et al.*, 2006; Søvik and Kløve, 2007; Ström *et al.*, 2007). Further, N_2O consumption was reported in subtropical mangroves (Allen *et al.*, 2011), a hypertrophic boreal lake (Huttunen *et al.*, 2002), and ponds constructed for nitrogen removal pond treating nutrients from wastewater (Johansson *et al.*, 2003).

Finally, the finding that GHGs emitted from Sonso Lagoon were comparable to EWWT calls for efforts to alleviate the impact of anthropogenic activities on FWs. For instance, avoiding domestic wastewater sewage, improving wastewater treatment and controlling fertilizer can prevent GHG emissions from FWs. A trend to increase anthropogenic activities will affect the most relevant feature of wetlands: their ability to sequester and store GHGs.

7.5.2 Environmental factors influencing GHG emissions from a eutrophic tropical natural wetland

The spatial variability of CO_2, CH_4 and N_2O fluxes was explained by water characteristics of Sonso Lagoon in the different points monitored (Chapter 6). The Sonso Lagoon exchanges water and sediments with a river (P1), receives domestic sewage and agricultural runoff discharges (P2), and is further dominated by phytoplankton (P3) and water-floating plants (P4).

In Chapter 6, considering the Carlson Trophic State Index (TSI), it was found that the trophic state condition of Sonso Lagoon varies from mesotrophic to eutrophic; thus, points P1 and P2 were in the mesotrophic state, and points P3 and P4 were in the eutrophic state. The higher TSI index at P3 and P4 could be explained by the direct discharge of domestic wastewater into Sonso Lagoon, which causes an increase in N and P in the water.

In Chapter 6, the GHG production was found to be mainly determined by organic matter (COD), nutrients (NO_3^-, NH_4^+, and TP) and environmental factors such as DO, pH, and temperature (tables 6.2, 6.3, and 6.4). The increase in organic matter and the reduction in DO levels can be considered as the key factors related to the emission of CO_2 and CH_4 from Sonso Lagoon. In wetlands, CO_2 and CH_4 emissions are produced by the decomposition of organic matter accumulated in the sediments (Yang *et al.*, 2014). Further, low DO levels favor anaerobic decomposition of organic matter leading to CO_2 and CH_4 production (Bridgham *et al.*, 2013). Thus, CO_2 and CH_4 were emitted mainly in the points P1, P2 and P3 that were characterized by high pollution due to domestic sewage discharge and agricultural run-off (figs. 6.1, 6.2 and 6.3 in Chapter 6).

The positive correlation between NO_3^- and methane production found in this study has also been reported in previous studies (Johansson *et al.*, 2004; Søvik and Kløve, 2007). Nitrates serve as the first terminal electron acceptor in wetland soils after oxygen depletion, making them an important compound in the oxidation of organic matter in wetlands (Sánchez-Carrillo *et al.*, 2011) to generate CO_2 and CH_4. In the same way, the results indicate that an increase in NH_4^+ increased CH_4 fluxes from Sonso Lagoon. This increase may be explained because elevated NH_4^+ concentration inhibits CH_4 oxidation (Biswas *et al.*, 2007; Borrel *et al.*, 2011). This inhibition is attributed to competition between NH_4 and CH_4 for binding sites on methane monooxygenase, because of their similar chemical structure (Bédard and Knowles, 1989).

Nitrous oxide is an intermediate of both nitrification and denitrification and its atmospheric release depends on the availability of N (NO_3^-, NO_2^-, or NH_4^+) and oxygen (Groffman, 1991; Morris, 1991). The presence of NO_3, NO_2, or NH_4^+ in Sonso Lagoon is related to wastewater discharge, sediment exchange from the Cauca River, and agricultural run-off. Nitrifying bacteria may produce NO and N_2O either as a side-product in the catabolic pathway (oxidizing ammonia to nitrite), or, alternatively, denitrifying bacteria may produce NO or N_2O converting nitrite with ammonia, hydrogen or pyruvate as an electron donor (Colliver and Stephenson, 2000; Wrage *et al.*, 2001; Law *et al.*, 2012). Nevertheless, it is difficult to conclude what the main processes are for N_2O production in the present study as nitrification and denitrification rates were not estimated.

The presence of PO_4^{-3} and NH_4^+ was positively correlated to CH_4 fluxes while apparently higher NO_3 coincided with lower CH_4 fluxes. Higher PO_4^{-3} concentrations coincided with higher CH_4 emissions measured in Sonso Lagoon, as was reported for lakes and ditches from The Netherlands and Sweden (Johansson et al., 2004; Schrier-Uijl et al., 2011). An increment of PO_4^{-3} in lakes under eutrophication stimulates the transformation of organic matter into methane (Adhya et al., 1998; Huttunen et al., 2003; Conrad and Klose, 2005; Sun et al., 2013).

Chapter 6 revealed that the eutrophic state in Sonso Lagoon affected its ability to sequester GHGs. As long as anthropogenic activities grow significantly, FW increases its organic matter and nutrient concentration i.e. N, P. This results in severe alterations of the water quality and function of these ecosystems i.e. eutrophication (Ventura, 2014). Eutrophication in FWs generally promotes excessive algae and plant growth and decay, favoring simple algae and plankton over other more complicated plants, and this can cause severe reduction in water quality as a result of oxygen depletion (Søndergaard, 2007). In addition, the nutrient loading enhances organic matter decomposition and microbial activity (Wright et al., 2009), which may lead to increased accumulation of carbon and nitrogen (Huttunen et al., 2001).

As a result, eutrophication affects the freshwater wetland biogeochemistry, leading to acceleration of the exchange of greenhouse gases between freshwater wetlands and the atmosphere (Casper et al., 2000; Huttunen et al., 2001).

7.5.3 Influence of floating aquatic macrophytes on GHG emissions from eutrophic tropical natural wetland

The negative fluxes observed in Sonso Lagoon occurred mainly in zones dominated by floating aquatic plants (Chapter 6). The capacity of floating macrophytes to contribute to sequester GHGs is based on: (i) acting as a physical barrier to prevent C diffusion across the water interface into the atmosphere (Van der Steen et al., 2003; Silva et al., 2012); (ii) sequestering CO_2 by algae and aquatic-floating plant photosynthesis (Brix et al., 2001; Teiter and Mander, 2005; Ström et al., 2007); (iii) favouring methane oxidation through translocation of oxygen gas produced by photosynthetic activity of the green leaves to the stems and roots and to the water body (Laanbroek, 2010) and (iv) the presence of attached methylotrophs in biofilms attached to floating leaves (Whalen, 2005; Chowdhury and Dick, 2013; Wang et al., 2013).

The GHG consumptions observed in Sonso Lagoon corresponds well with that reported in the literature(Brix et al., 2001; Johansson et al., 2003; Longhi et al., 2008; Attermeyer et al., 2016). However, other studies have emphasized that FWs dominated by emergent macrophytes emit more GHGs (Duan et al., 2005; Bergström et al., 2007)

Although the consumption of N_2O has been reported in other studies (Wang *et al.*, 2016), it remains unclear which one or which are the mechanisms that explain this phenomenon. A first hypothesis is that probably the vegetation cover avoided the exchange of this gas into the atmosphere. A second hypothesis is that this gas is consumed by denitrifying bacteria (Johansson *et al.*, 2003). However, these processes were not quantified in this study and therefore this will be a subject for further studies to elucidate the extension of these mechanisms on GHG emissions from wetlands.

7.6 EMISSIONS IN CO₂ EQUIVALENTS FOR THE EWWT STUDIED

In order to compare the impacts of different EWWT, a metric, the global warming potential (GWP) is used (IPCC, 2014). Based on GWP, the emissions of CH_4 and N_2O are transformed into the equivalent emissions of CO_2, multiplied by 28 and 265, respectively. The IPCC (2014) excludes CO_2 emission in the greenhouse gas emission inventory for wastewater treatment systems (WWTS), because the IPCC recognizes that CO_2 emission from WWTS is from biogenic origins and it does not contribute to the increase of anthropogenic CO_2 concentrations in the atmosphere (Shaw and Koh, 2014). However, CO_2 contribution is included in **Table 7.2** only to give a perspective of the emission magnitude.

Table 7.2 Global warming potential (GWP) in gCO_{2eq} m^{-2} d^{-1} for the EWWT studied. Calculated from CO_2, CH_4, and N_2O median fluxes.

EWWT type		CH₄ emission (gCO_{2eq} m^{-2} d^{-1})	N₂O emission (gCO_{2eq} m^{-2} d^{-1})	Total (gCO_{2eq} m^{-2} d^{-1})
AP	98.2	1390	3	1491.2
AFP pilot-scale	1.9	2.1	1.6	5.6
AFP full-scale	0.9	66.1	0.8	67.8
DBP pilot-scale	1.7	5.0	1.4	8.1
Sonso Lagoon	1.4	9.8	0	11.2

As was expected, the highest GWP corresponds to AP investigated in Chapter 3 being attributable to methane emissions. The ratio CH_4/N_2O suggests that 99.7% of GWP was related to CH_4 emitted from AP (excluding CO_2 emissions). On the other hand, in SFPs the values of GWP were significantly different when compared with pilot and full-scale systems. The GWP index from the full-scale AFP was higher than both AFP and DBP pilot-scale. This can be explained because the largest CH_4 emissions from the full-scale AFP amounted to 99% of GWP. As was mentioned, the full-scale AFP was operated under a higher organic load and additionally was more depth than pilot-scale systems favouring methane production. In addition, the pilot-scale AFP showed a lower GWP than the pilot-scale DBP. Surprisingly, the ratio CH_4/N_2O in the pilot-scale AFP showed only a 57% of contribution of methane to GWP, the most important being the N_2O contribution. As was discussed, the N_2O emitted from the pilot-scale was almost twofold that measured in the full-scale AFP.

Based on the organic matter removed and using the GWP shown in **Table 7.2**, it was found that the CO_{2-eq} in the EWWT studied was between 1.5-2.18 $kgCO_{2-eq}.kgCOD_{removed}^{-1}$. These values were higher than reported in conventional aerobic/anaerobic full treatment systems that were between 0.91 and 1.04 kg $CO_{2eq}.kgCOD_{removed}^{-1}$ (Keller and Hartley, 2003; Flores-Alsina *et al.*, 2011). These GWP values indicate the large potential of WSPs to contribute to global warming and address the necessity to improve these systems from the perspective of preventing their GHG emissions.

An estimation of the WSPs contribution to the national inventory of GHGs in Colombia can be carried out using the GWP calculated (Table 7.2). In Colombia only 9% of wastewater is treated and of this total 55% are treated in wastewater stabilization ponds. Based on these figures, a rough estimation shows that 1,807 tons of $CO_{2eq}.d^{-1}$ (0.66 $MtCO_{2eq}.yr^{-1}$) can be emitted from WSPs in Colombia. This value is equivalent to 0.3% of the total GHG emissions reported for Colombia of 180 $MtCO_{2eq}.yr^{-1}$ (IDEAM, 2008). This contribution could be much greater with the projected increase of wastewater treatment coverage, for instance, if over 90% of wastewater was treated then the contribution by WSPs in Colombia would amount to 3%.

The results from Chapter 6 demonstrate that tropical freshwater wetlands may be particularly important GHG sources to the atmosphere. In general, the magnitudes of CO_2 measured in Sonso Lagoon were higher than quantified for CH_4 and N_2O. However, given the greater global warming potential of CH_4 and N_2O, these emissions are important from the perspective of the contribution to radiative force.

7.7 OUTLOOK

In this research, the GHG fluxes from EWWT under tropical conditions have been measured and the main factors influencing these emissions have also been identified. In this context, it is evident that further studies should be primarily focused on:

* GHG measurements using static chambers should be combined with approaches based on Eddy Covariance to compare and cross-validate fluxes estimated from EWWTs. This can provide a best resolution and accuracy determining the spatial variation of GHGs in EWWTs. Further, it is strongly recommended to quantify the ebullition of GHGs from EWWTs because in some EWWTs there can be an important mechanism contributing to emissions.

- The mechanisms that contributed to N_2O emissions in anaerobic ponds should be elucidated. In this thesis, a relatively low N_2O flux from the AP was identified, but this did not correlate to environmental factors operating in the pond. This raises the question: what is the pathway in an AP that regulates N_2O production?

- As was demonstrated, open WSPs contribute significantly to global warming. This shows the need to develop new research on the way forward to recovering biogas from AP. Because a high proportion of methane is dissolved in the effluent of an anaerobic pond, there is the necessity to recover this methane before its discharge to SFPs. Probably the stripping of this gas followed by purification may be an alternative for reducing CH_4 emissions from APs.

- In this thesis, the changes in GHG emissions from SFPs due to environmental conditions and water characteristics were observed. However, there is a demand for further studies to understand better the pathways of GHG production in EWWTs. For future work, it is suggested to investigate CH_4 oxidation in the water column and sediments. Further, the understanding of CO_2, CH_4 and N_2O also require measurements of the concentrations of these gases in the whole water column to determine the transport mechanism into atmosphere of these gases. On the other hand, a gap in the knowledge is related to the mechanisms that favour N_2O consumption in EWWTs.

- An identification of the algae type and its physiology in AFPs is necessary to elucidate the capacity of different algae to take up CO_2.

- This research suggests that floating aquatic macrophytes can act as a sink for GHGs in tropical freshwater wetland. This aspect calls for additional studies to better understand how aquatic macrophytes affect local, regional, and global carbon cycling.

- The study in Sonso Lagoon was focused on the measurement of diffusive flux through the water surface; however, FWs are complex and dynamic systems where many equilibriums and processes occur simultaneously in the water column and sediments. This is important to examine because the different forcing functions such as pH, temperature, substrates concentration influence this dynamic. Besides, the internal cycling of C and N either in the water column or in the flooded soils and sediments has to be properly understood in FWs. Thus, these factors should be studied through lab and field studies.

- In developing countries there are many uncertainties with respect to direct emissions, indirect emissions and availability and quality of data about GHGs from the wastewater sector (Bogner *et al.*, 2007). In order to reduce the gap and to gain knowledge about reliable national data, a field measurement campaign for each country or region should be carried out. The purpose of this is to understand and control the emission of CH_4 and N_2O from wastewater collection and treatment systems.

7.8 CONCLUSIONS

- This study has provided one of the very first sets of comprehensive CO_2, CH_4, and N_2O emission results for EWWT under tropical conditions. According to the results, all the EWWT under the studied conditions were net sources of CO_2 CH_4, and N_2O when taking into account a 24-hr period. Thus, this study provides evidence that EWWT impact the environment from the point of view of GHG emissions, contributing to global warming.

- The results of this study showed that in many cases the data obtained using static chambers show a non-linear trend. Therefore, when fluxes are estimated assuming only linear conditions there will be an overestimation in the calculation of GHG emissions. In general, the underestimation ranged between 10 and 50 % on emissions. Thus, it is strongly recommended that the data be adjusted to non-linear models and for its evaluation the R^2_{adj} criterion is used, because it detects the influence of extra parameters in the regression models, decreasing the uncertainties in the flux calculation.

- Biogas recovery from municipal wastewater through anaerobic processes might not be economically practical or environmentally friendly due to the fact that a substantial amount of methane, around 20%, is dissolved into the treated effluent. In addition, low operation and maintenance i.e. sludge accumulation reduces biogas production. Because the collection of biogas for energy production in open APs is limited, the release of CO_2 and CH_4 from the anaerobic pond surface is contributing to global warming.

- It is clear that tropical conditions have an influence on GHGs from EWWT. High temperatures stimulate the biochemical process leading to GHG production. In APs a high temperature favours anaerobic decomposition, and CO_2 and CH_4, production is increased. Also, the photosynthesis is favoured by a long photoperiod which contributes to the GHG dynamics in EWWT. The differences found between daytime and night-time CO_2, CH_4, and N_2O fluxes suggest that the GHG dynamic in FSPs is influenced strongly by the photoperiod. A higher photosynthesis activity increases CO_2 uptake by algae and floating macrophytes. Moreover, photosynthesis influences the pH and DO affecting the CH_4 and N_2O production. This is explained in part because GHG emissions measured in the EWWT studied under tropical conditions were higher than reported under other climatic condition i.e. Subtropical.

- When estimating the CO_2, CH_4, and N_2O fluxes from SFPs, the fluxes should be measured considering the daytime and night-time measurements. Overall, when photosynthesis is omitted the fluxes from the SFPs are probably overestimated and the consequence of this is a large uncertainty in GHG estimation from the SFPs.

- The GHG fluxes from Sonso Lagoon were the same order of magnitude as those reported in wastewater treatment such as constructed wetlands and facultative ponds. This calls for efforts to alleviate the impact of anthropogenic activities on FWs. For instance, avoiding domestic wastewater sewage, improving wastewater treatment and controlling fertilizer can prevent GHG emissions from FWs. A trend to increase anthropogenic activities affects the most relevant feature of wetlands, i.e. their ability to sequester and store GHGs.

- This study demonstrates that floating aquatic macrophytes play a role in the dynamic of GHGs in freshwater wetlands. Floating aquatic plants contribute to sequestering GHGs and can affect the balance of net C and N emissions from FWs. Therefore, in the global wetland models about GHG estimation, the contribution of floating aquatic macrophytes should be considered. Of course, this involves an effort to plug gaps in understanding plant physiology, carbon inputs from roots, and autotrophic and heterotrophic respiration related to photosynthesis.

7.9 REFERENCES

Adhya, T., Pattnaik, P., Satpathy, S., Kumaraswamy, S. and Sethunathan, N. (1998). Influence of phosphorus application on methane emission and production in flooded paddy soils. Soil Biology and Biochemistry 30(2), 177-181.

Allen, D., Dalal, R., Rennenberg, H. and Schmidt, S. (2011). Seasonal variation in nitrous oxide and methane emissions from subtropical estuary and coastal mangrove sediments, Australia. Plant Biology 13(1), 126-133.

Attermeyer, K., Flury, S., Jayakumar, R., Fiener, P., Steger, K., Arya, V., Wilken, F., Van Geldern, R. and Premke, K. (2016). Invasive floating macrophytes reduce greenhouse gas emissions from a small tropical lake. Scientific reports 6, 20424.

Bédard, C. and Knowles, R. (1989). Physiology, biochemistry, and specific inhibitors of CH_4, NH_4^+, and CO oxidation by methanotrophs and nitrifiers. Microbiological reviews 53(1), 68-84.

Bergström, I., Mäkelä, S., Kankaala, P. and Kortelainen, P. (2007). Methane efflux from littoral vegetation stands of southern boreal lakes: an upscaled regional estimate. Atmospheric Environment 41(2), 339-351.

Biswas, H., Mukhopadhyay, S.K., Sen, S. and Jana, T.K. (2007). Spatial and temporal patterns of methane dynamics in the tropical mangrove dominated estuary, NE coast of Bay of Bengal, India. Journal of Marine Systems 68(1–2), 55-64.

Bogner, J., Abderalfie, A., Diaz, C., Faaij, A., Gao, Q., Hashimoto, S., Mareckova, K., Pipatti, R. and Zhang, T. (2007). Waste Management, In Climate Change 2007: Mitigation. Contribution of Working Group III to the Fourth Assessment Report of the

Intergovernmental Panel on Climate Change. B. Metz, O.R. Davidson, P.R. Bosch, R. Dave and Meyer, L.A. (eds), p. 32, Cambridge University, Cambridge, United Kingdom

Borrel, G., Jézéquel, D., Biderre-Petit, C., Morel-Desrosiers, N., Morel, J.-P., Peyret, P., Fonty, G. and Lehours, A.-C. (2011). Production and consumption of methane in freshwater lake ecosystems. Research in Microbiology 162(9), 832-847.

Bridgham, S.D., Cadillo-Quiroz, H., Keller, J.K. and Zhuang, Q. (2013). Methane emissions from wetlands: biogeochemical, microbial, and modeling perspectives from local to global scales. Global Change Biology 19(5), 1325-1346.

Brix, H., Sorrell, B.K. and Lorenzen, B. (2001). Are Phragmites-dominated wetlands a net source or net sink of greenhouse gases? Aquatic Botany 69(2-4), 313-324.

Casper, P., Maberly, S.C., Hall, G.H. and Finlay, B.J. (2000). Fluxes of methane and carbon dioxide from a small productive lake to the atmosphere. Biogeochemistry 49(1), 1-19.

Chowdhury, T.R. and Dick, R.P. (2013). Ecology of aerobic methanotrophs in controlling methane fluxes from wetlands. Applied soil ecology 65, 8-22.

Ciais, P., Sabine, C., Bala, G., Bopp, L., Brovkin, V., Canadell, J., Chhabra, A., DeFries, R., Galloway, J. and Heimann, M. (2014). Climate change 2013: the physical science basis. Contribution of Working Group I to the Fifth Assessment Report of the Intergovernmental Panel on Climate Change, pp. 465-570, Cambridge University Press.

Colliver, B. and Stephenson, T. (2000). Production of nitrogen oxide and dinitrogen oxide by autotrophic nitrifiers. Biotechnology Advances 18(3), 219-232.

Conrad, R. and Klose, M. (2005). Effect of potassium phosphate fertilization on production and emission of methane and its ^{13}C-stable isotope composition in rice microcosms. Soil Biology and Biochemistry 37(11), 2099-2108.

Crill, P.M., Bartlett, K.B., Harriss, R.C., Gorham, E., Verry, E.S., Sebacher, D.I., Madzar, L. and Sanner, W. (1988). Methane flux from Minnesota peatlands. Global biogeochemical cycles 2(4), 371-384.

Daelman, M., van Voorthuizen, E.M., Van Dongen, L., Volcke, E. and Van Loosdrecht, M. (2013). Methane and nitrous oxide emissions from municipal wastewater treatment– results from a long-term study. Water Science and Technology 67(10), 2350-2355.

Daelman, M.R., van Voorthuizen, E.M., van Dongen, U.G., Volcke, E.I. and van Loosdrecht, M.C. (2015). Seasonal and diurnal variability of N2O emissions from a full-scale municipal wastewater treatment plant. Science of the Total Environment 536, 1-11.

De Klein, J.J. and Van der Werf, A.K. (2014). Balancing carbon sequestration and GHG emissions in a constructed wetland. Ecological Engineering 66, 36-42.

Detweiler, A.M., Bebout, B.M., Frisbee, A.E., Kelley, C.A., Chanton, J.P. and Prufert-Bebout, L.E. (2014). Characterization of methane flux from photosynthetic oxidation ponds in a wastewater treatment plant. Water Science & Technology 70, 980-989.

Duan, X., Wang, X., Mu, Y. and Ouyang, Z. (2005). Seasonal and diurnal variations in methane emissions from Wuliangsu Lake in arid regions of China. Atmospheric Environment 39(25), 4479-4487.

Fey, A., Benckiser, G. and Ottow, J.C.G. (1999). Emissions of nitrous oxide from a constructed wetland using a groundfilter and macrophytes in waste-water purification of a dairy farm. Biology and Fertility of Soils 29, 354-359.

Flores-Alsina, X., Corominas, L., Snip, L. and Vanrolleghem, P.A. (2011). Including greenhouse gas emissions during benchmarking of wastewater treatment plant control strategies. Water Research 45(16), 4700-4710.

Glaz, P., Bartosiewicz, M., Laurion, I., Reichwaldt, E.S., Maranger, R. and Ghadouani, A. (2016). Greenhouse gas emissions from waste stabilisation ponds in Western Australia and Quebec (Canada). Water Research 101, 64-74.

Groffman, P.M. (1991). Ecology of nitrification and denitrification in soil evaluated at scales relevant to atmospheric chemistry. Rogers, J. and Whitman, W. (eds) Microbial production and consumption of greenhouse gases: methane, nitrogen oxides, and halomethanes, American Society of Biology, Washington DC, 201-217.

Gui, P., Inamori, R., Matsumura, M. and Inamori, Y. (2007). Evaluation of constructed wetlands by wastewater purification ability and greenhouse gas emissions. Water Science & Technology 56, 49-55.

Guo, G., Wang, Y., Hao, T., Wu, D. and Chen, G.-H. (2018). Enzymatic nitrous oxide emissions from wastewater treatment. Frontiers of Environmental Science & Engineering 12(1), 10.

Hernandez-Paniagua, I.Y., Ramirez-Vargas, R., Ramos-Gomez, M.S., Dendooven, L., Avelar-Gonzalez, F.J. and Thalasso, F. (2014). Greenhouse gas emissions from stabilization ponds in subtropical climate. Environmental Technology 35, 727-734.

Hopfensperger, K.N., Gault, C.M. and Groffman, P.M. (2009). Influence of plant communities and soil properties on trace gas fluxes in riparian northern hardwood forests. Forest Ecology and Management 258(9), 2076-2082.

Huttunen, J.T., Alm, J., Liikanen, A., Juutinen, S., Larmola, T., Hammar, T., Silvola, J. and Martikainen, P.J. (2003). Fluxes of methane, carbon dioxide and nitrous oxide in boreal lakes and potential anthropogenic effects on the aquatic greenhouse gas emissions. Chemosphere 52(3), 609-621.

Huttunen, J.T., Hammar, T., Alm, J., Silvola, J. and Martikainen, P.J. (2001). Greenhouse gases in non-oxygenated and artificially oxygenated eutrophied lakes during winter stratification. Journal of environmental quality 30(2), 387-394.

Huttunen, J.T., Väisänen, T.S., Heikkinen, M., Hellsten, S., Nykänen, H., Nenonen, O. and Martikainen, P.J. (2002). Exchange of CO_2, CH_4 and N_2O between the atmosphere and two northern boreal ponds with catchments dominated by peatlands or forests. Plant and Soil 242(1), 137-146.

IDEAM (2008). Segunda Comunicación Nacional de Colombia ante la Convención Marco de las Naciones Unidas sobre Cambio Climático. IDEAM (ed), IDEAM, Bogotá.

Itokawa, H., Hanaki, K. and Matsuo, T. (2001). Nitrous oxide production in high-loading biological nitrogen removal process under low COD/N ratio condition. Water Research 35(3), 657-664.

Jauhiainen, J., Hooijer, A. and Page, S. (2012). Carbon dioxide emissions from an Acacia plantation on peatland in Sumatra, Indonesia. Biogeosciences 9(2), 617-630.

Jauhiainen, J., Takahashi, H., Heikkinen, J.E., Martikainen, P.J. and Vasander, H. (2005). Carbon fluxes from a tropical peat swamp forest floor. Global Change Biology 11(10), 1788-1797.

Johansson, A., Gustavsson, A.-M., Öquist, M. and Svensson, B. (2004). Methane emissions from a constructed wetland treating wastewater—seasonal and spatial distribution and dependence on edaphic factors. Water Research 38(18), 3960-3970.

Johansson, A., Klemedtsson, Å.K., Klemedtsson, L. and Svensson, B. (2003). Nitrous oxide exchanges with the atmosphere of a constructed wetland treating wastewater. Tellus B 55(3), 737-750.

Kayombo, S., Mbwette, T., Mayo, A., Katima, J. and Jørgensen, S. (2002). Diurnal cycles of variation of physical–chemical parameters in waste stabilization ponds. Ecological Engineering 18(3), 287-291.

Keller, J. and Hartley, K. (2003). Greenhouse gas production in wastewater treatment: process selection is the major factor. Water Science & Technology 47(12), 43-48.

Konaté, Y., Maiga, A.H., Casellas, C. and Picot, B. (2013). Biogas production from an anaerobic pond treating domestic wastewater in Burkina Faso. Desalination and Water Treatment 51(10-12), 2445-2452.

Krithika, K., Purvaja, R. and Ramesh, R. (2008). Fluxes of methane and nitrous oxide from an Indian mangrove. Current Science, 218-224.

Laanbroek, H.J. (2010). Methane emission from natural wetlands: interplay between emergent macrophytes and soil microbial processes. A mini-review. Annals of Botany 105(1), 141-153.

Law, Y., Lant, P. and Yuan, Z. (2011). The effect of pH on N2O production under aerobic conditions in a partial nitritation system. Water Research 45(18), 5934-5944.

Law, Y., Ye, L., Pan, Y. and Yuan, Z. (2012). Nitrous oxide emissions from wastewater treatment processes. Philosophical Transactions of the Royal Society B: Biological Sciences 367(1593), 1265-1277.

Liikanen, A., Huttunen, J.T., Karjalainen, S.M., Heikkinen, K., Vaisanen, T.S., Nykanen, H. and Martikainen, P.J. (2006). Temporal and seasonal changes in greenhouse gas emissions from a constructed wetland purifying peat mining runoff waters. Ecological Engineering 26(3), 241-251.

Longhi, D., Bartoli, M. and Viaroli, P. (2008). Decomposition of four macrophytes in wetland sediments: Organic matter and nutrient decay and associated benthic processes. Aquatic Botany 89(3), 303-310.

Mander, U., Teiter, S. and Augustin, J. (2005). Emission of greenhouse gases from constructed wetlands for wastewater treatment and from riparian buffer zones. Water Sci Technol 52(10-11), 167-176.

Melling, L., Hatano, R. and Goh, K.J. (2005). Methane fluxes from three ecosystems in tropical peatland of Sarawak, Malaysia. Soil Biology and Biochemistry 37(8), 1445-1453.

Metcalf, L. and Eddy, H. (2003) Wastewater Engineering: Treatment, Disposal and Reuse, , Mc Graw Hill, New York.

Morris, J.T. (1991). Effects of nitrogen loading on wetland ecosystems with particular reference to atmospheric deposition. Annual Review of Ecology and Systematics 22(1), 257-279.

Nahlik, A.M. and Mitsch, W.J. (2011). Methane emissions from tropical freshwater wetlands located in different climatic zones of Costa Rica. Global Change Biology 17(3), 1321-1334.

Page, S.E., Rieley, J.O. and Banks, C.J. (2011). Global and regional importance of the tropical peatland carbon pool. Global Change Biology 17(2), 798-818.

Paredes, M.G., Güereca, L.P., Molina, L.T. and Noyola, A. (2015). Methane emissions from stabilization ponds for municipal wastewater treatment in Mexico. Journal of Integrative Environmental Sciences, 1-15.

Pedersen, A.R., Petersen, S.O. and Schelde, K. (2010). A comprehensive approach to soil-atmosphere trace-gas flux estimation with static chambers. European Journal of Soil Science 61(6), 888-902.

Picot, B., Paing, J., Sambuco, J.P., Costa, R.H.R. and Rambaud, A. (2003). Biogas production, sludge accumulation and mass balance of carbon in anaerobic ponds. Water Science and Technology 48(2), 243-250.

Sánchez-Carrillo, S., Angeler, D., Álvarez-Cobelas, M. and Sánchez-Andrés, R. (2011). Eutrophication: causes, consequences and control, pp. 195-210, Springer.

Saunois, M., Bousquet, P., Poulter, B., Peregon, A., Ciais, P., Canadell, J.G., Dlugokencky, E.J., Etiope, G., Bastviken, D., Houweling, S., Janssens-Maenhout, G., Tubiello, F.N., Castaldi, S., Jackson, R.B., Alexe, M., Arora, V.K., Beerling, D.J., Bergamaschi, P., Blake, D.R., Brailsford, G., Brovkin, V., Bruhwiler, L., Crevoisier, C., Crill, P., Covey, K., Curry, C., Frankenberg, C., Gedney, N., Hoglund-Isaksson, L., Ishizawa, M., Ito, A., Joos, F., Kim, H.S., Kleinen, T., Krummel, P., Lamarque, J.F., Langenfelds, R., Locatelli, R., Machida, T., Maksyutov, S., McDonald, K.C., Marshall, J., Melton, J.R., Morino, I., Naik, V., O'Doherty, S., Parmentier, F.J.W., Patra, P.K., Peng, C.H., Peng, S.S., Peters, G.P., Pison, I., Prigent, C., Prinn, R., Ramonet, M., Riley, W.J., Saito, M., Santini, M., Schroeder, R., Simpson, I.J., Spahni, R., Steele, P., Takizawa, A., Thornton, B.F., Tian, H.Q., Tohjima, Y., Viovy, N., Voulgarakis, A., van Weele, M., van der Werf, G.R., Weiss, R., Wiedinmyer, C., Wilton, D.J., Wiltshire, A., Worthy, D., Wunch, D., Xu, X.Y., Yoshida, Y., Zhang, B., Zhang, Z. and Zhu, Q. (2016). The global methane budget 2000-2012. Earth System Science Data 8(2), 697-751.

Schrier-Uijl, A., Veraart, A., Leffelaar, P., Berendse, F. and Veenendaal, E. (2011). Release of CO_2 and CH_4 from lakes and drainage ditches in temperate wetlands. Biogeochemistry 102(1), 265-279.

Silva, J.P., Ruiz, J.L., Peña, M.R., Lubberding, H. and Gijzen, H. (2012). Influence of photoperiod on carbon dioxide and methane emissions from two pilot-scale stabilization ponds. Water Science & Technology 66(9).

Singh, V.P., Dass, P., Kaur, K., Billore, S.K., Gupta, P.K. and Parashar, D.C. (2005). Nitrous oxide fluxes in a tropical shallow urban pond under influencing factors. Current Science 88(3), 478.

Sjögersten, S., Black, C.R., Evers, S., Hoyos-Santillan, J., Wright, E.L. and Turner, B.L. (2014). Tropical wetlands: A missing link in the global carbon cycle? Global biogeochemical cycles 28(12), 1371-1386.

Søndergaard, M. (2007). Nutrient dynamics in lakes-with emphasis on phosphorus, sediment and lake restoration, Aarhus Universitet, Danmarks Miljøundersøgelser, Afdeling for Ferskvandsøkologi.

Song, C., Zhang, L., Wang, Y. and Zhao, Z. (2006). Annual dynamics of CO_2, CH_4, N_2O emissions from freshwater marshes and affected by nitrogen fertilization. Huan jing ke xue 27(12), 2369-2375.

Søvik, A. and Kløve, B. (2007). Emission of N_2O and CH_4 from a constructed wetland in southeastern Norway. Science of the Total Environment 380(1), 28-37.

Sovik, A.K., Augustin, J., Heikkinen, K., Huttunen, J.T., Necki, J.M., Karjalainen, S.M., Klove, B., Liikanen, A., Mander, U. and Puustinen, M. (2006). Emission of the Greenhouse Gases Nitrous Oxide and Methane from Constructed Wetlands in Europe. Journal of environmental quality 35(6), 2360.

Stadmark, J. and Leonardson, L. (2005). Emissions of greenhouse gases from ponds constructed for nitrogen removal. Ecological Engineering 25(5), 542-551.

Ström, L., Lamppa, A. and Christensen, T.R. (2007). Greenhouse gas emissions from a constructed wetland in southern Sweden. Wetlands Ecology and Management 15(1), 43-50.

Sun, Q.-Q., Shi, K., Damerell, P., Whitham, C., Yu, G.-H. and Zou, C.-L. (2013). Carbon dioxide and methane fluxes: Seasonal dynamics from inland riparian ecosystems, northeast China. Science of the Total Environment 465, 48-55.

Suzuki, I., Kwok, S.-C. and Dular, U. (1976). Competitive inhibition of ammonia oxidation in Nitrosomonas europaea by methane, carbon monoxide or methanol. FEBS letters 72(1), 117-120.

Tanner, C.C., Adams, D.D. and Downes, M.T. (1997). Methane emissions from constructed wetlands treating agricultural wastewaters. Journal of environmental quality 26(4), 1056-1062.

Teiter, S. and Mander, U. (2005). Emission of N_2O, N_2, CH_4, and CO_2 from constructed wetlands for wastewater treatment and from riparian buffer zones. Ecological Engineering 25(5), 528-541.

Toprak, H.k. (1995). Temperature and organic loading dependency of methane and carbon dioxide emission rates of a full-scale anaerobic waste stabilization pond. Water Research 29(4), 1111-1119.

Tremblay, A., Lambert, M. and Gagnon, L. (2004). Do hydroelectric reservoirs emit greenhouse gases? Environmental Management 33, 509-517.

Vadivelu, V.M., Keller, J. and Yuan, Z. (2007). Effect of free ammonia on the respiration and growth processes of an enriched Nitrobacter culture. Water Research 41(4), 826-834.

Vadivelu, V.M., Yuan, Z., Fux, C. and Keller, J. (2006). The inhibitory effects of free nitrous acid on the energy generation and growth processes of an enriched Nitrobacter culture. Environmental Science & Technology 40(14), 4442-4448.

Van der Steen, N.P., Ferrer, A.V.M., Samarasinghe, K.G. and Gijzen, H.J. (2003). Quantification and comparison of methane emissions from algae and duckweed based wastewater treatment ponds. Universidad del Valle; CINARA; International Water Association. Memorias del evento: Agua 2003. Cartagena de Indias, IWA, 2003, p. 1-7.

Ventura, R.E. (2014). Wetlands and Greenhouse Gas Fluxes: Causes and Effects of Climate Change–A Meta-Analysis, Pomona College.

Vieira, A., Galinha, C., Oehmen, A. and Carvalho, G. (2018). The link between nitrous oxide emissions, microbial community profile and function from three full-scale WWTPs. Science of the Total Environment.

Wang, H., Liao, G., D'Souza, M., Yu, X., Yang, J., Yang, X. and Zheng, T. (2016). Temporal and spatial variations of greenhouse gas fluxes from a tidal mangrove wetland in Southeast China. Environmental Science and Pollution Research 23(2), 1873-1885.

Wang, J., Zhang, J., Xie, H., Qi, P., Ren, Y. and Hu, Z. (2011). Methane emissions from a full-scale A/A/O wastewater treatment plant. Bioresource technology 102(9), 5479-5485.

Wang, Y., Yang, H., Ye, C., Chen, X., Xie, B., Huang, C., Zhang, J. and Xu, M. (2013). Effects of plant species on soil microbial processes and CH_4 emission from constructed wetlands. Environmental pollution 174, 273-278.

Whalen, S. (2005). Biogeochemistry of methane exchange between natural wetlands and the atmosphere. Environmental Engineering Science 22(1), 73-94.

Wrage, N., Velthof, G., Van Beusichem, M. and Oenema, O. (2001). Role of nitrifier denitrification in the production of nitrous oxide. Soil Biology and Biochemistry 33(12), 1723-1732.

Wright, A.L., Ramesh Reddy, K.R. and Newman, S. (2009). Microbial indicators of eutrophication in Everglades wetlands. Soil Science Society of America Journal 73(5), 1597-1603.

Wright, E.L., Black, C.R., Turner, B.L. and Sjögersten, S. (2013). Environmental controls of temporal and spatial variability in CO_2 and CH_4 fluxes in a neotropical peatland. Global Change Biology 19(12), 3775-3789.

Wu, H., Lin, L., Zhang, J., Guo, W., Liang, S. and Liu, H. (2016). Purification ability and carbon dioxide flux from surface flow constructed wetlands treating sewage treatment plant effluent. Bioresource technology 219, 768-772.

Xing, Y., Xie, P., Yang, H., Ni, L., Wang, Y. and Rong, K. (2005). Methane and carbon dioxide fluxes from a shallow hypereutrophic subtropical Lake in China. Atmospheric Environment 39(30), 5532-5540.

Yang, J., Liu, J., Hu, X., Li, X., Wang, Y. and Li, H. (2013). Effect of water table level on CO_2, CH_4 and N_2O emissions in a freshwater marsh of Northeast China. Soil Biology and Biochemistry 61, 52-60.

Yang, L., Lu, F., Zhou, X., Wang, X., Duan, X. and Sun, B. (2014). Progress in the studies on the greenhouse gas emissions from reservoirs. Acta Ecologica Sinica 34(4), 204-212.

Summary

Nowadays there is worldwide concern about the impact of global warming and climate change on the environment and consequently on the people living on our planet. Excessive greenhouse gas emissions - mainly CO_2, CH_4 and N_2O - from anthropogenic sources are the main drivers for global warming and climate change. This is why the identification and quantification of all sources, both natural and anthropogenic, has become a key challenge for scientists and policy maker groups working on climate change or global warming, which is needed for developing strategies to control and reduce the rate of increase of the GHG emissions into the atmosphere.

Wastewater treatment systems have been used for the removal of organic carbon, nutrients and pathogenic microorganisms from wastewater. In wastewater treatment, organic matter and nutrients are removed when they are converted through microbial processes into gaseous compounds to be released into the atmosphere. However, some of these gases, such as CO_2, CH_4 and N_2O, can have adverse effects in the atmosphere because they act as greenhouse gases and cause global warming.

GHG emissions from the wastewater treatment sector represent 3-4% of total GHG emissions. However, the contribution of the wastewater sector to GHG emissions may be underestimated. This is due to large uncertainties with respect to direct emissions, indirect emissions, and the availability and quality of annual data for the wastewater sector. Global CH_4 and N_2O emissions from wastewater are estimated to have increased between 1990 and 2025 from 352 to 477 $MtCO_{2eq}$ and from 82 to 100 $MtCO_{2eq}$, respectively. This growth in GHG emissions come from developing countries in East and South Asia, the Middle East, the Caribbean, and Central and South America, mainly due to population increase. As long as populations grow significantly without large-scale advances in wastewater treatment, these areas will continue to have a major influence on the upward trend in GHG emissions.

Ecotechnologies for wastewater treatment (EWWT) combine ecological principles of natural systems with engineering principles to improve the removal of organic carbon, nutrients and pathogenic microorganisms from wastewater. The main EWWT recognized are anaerobic ponds, facultative ponds, duckweed-based ponds and constructed wetlands. EWWT are mainly solar-based systems, which makes the dependence on external energy minimal and therefore they are considered sustainable technologies. However, EWWT are an important source of GHG such as CO_2, CH_4 and N_2O and are probably contributing to global warming.

So far, limited information is available on the fate of GHG production or consumption in EWWT operated under tropical conditions. This PhD thesis quantifies the CO_2, CH_4 and N_2O fluxes generated in three EWWT such as anaerobic ponds, facultative ponds, and duckweed-based ponds. Likewise, it included the flux of GHGs from a freshwater wetland that is perturbed by anthropogenic activities such as wastewater discharge and nutrients from agricultural run-off. This research did not include constructed wetlands because, according to the literature review, these have received major attention regarding GHG measurements. Further, this thesis studied the influence of environmental factors such as temperature, pH, DO, and nutrients on GHG emissions produced in EWWT under tropical conditions. The methodology adopted in the research includes both pilot-scale and full-scale outdoor experiments. All the studies were performed using the wastewater produced in a small town located in Colombia called Ginebra and none of the studies used synthetic wastewater.

The first part of the research carried out in this thesis (Chapter 2) focused on the adaptation and assessment of a measurement technique to estimate the GHG flux produced in EWWT systems. The static chamber technique was chosen because it has been the most useful and reported technique in measuring GHG from wastewater plants, aquatic ecosystems and soils. The results obtained suggest that closed static chambers are a good analytical methodology to estimate GHG emissions from wastewater stabilization ponds. However, an intrinsic limitation of using closed static chambers was found which is that not all the data for gas concentrations measured within a chamber headspace can be used to estimate the flux due to gradient concentration curves with non-plausible and physical explanations. Based on the total data set (n = 47), the percentage of curves accepted were 93.6, 87.2, and 73% for CH_4, CO_2 and N_2O, respectively. In addition, the statistical analyses demonstrated that only taking linear regression into account was frequently inappropriate for the determination of GHG flux from stabilization ponds by the closed static chamber technique. In this work, it is clear that when $R^2_{adj-non-lin} > R^2_{adj-lin}$, the application of linear regression models is not recommended, as it leads to an underestimation of GHG fluxes by 10 to 50%. Therefore, the main conclusion of Chapter 2 is that incorrect use of the usual R^2 parameter and only the linear regression model to estimate the fluxes may lead to severe underestimation of the real contribution of GHG emissions from wastewater.

Undertaking a full-scale study on an anaerobic pond treating domestic wastewater was an attempt to provide performance data on GHG fluxes such as CO_2 and CH_4, and N_2O under tropical conditions (Chapter 3). CH_4 emissions ranged from 13.4 to 178.7 $L.m^{-2}.d^{-1}$, CO_2 from 9.3 to 130.5 $L.m^{-2}.d^{-1}$, while N_2O emissions ranged between 0.0016 and 0.013 $L.m^{-2}.d^{-1}$. According to the average fluxes the emission rates into the atmosphere for CH_4 and CO_2 were 0.24 m^3 CH_4/kg COD_{rem} and 0.18 m^3 CO_2/kg COD_{rem}, respectively. Further, a COD mass balance calculation indicated that 37% of the influent COD was converted to CH_4 and 36% left the anaerobic pond with the effluent. The rest of the COD was accounted for as volatile solids (3.5%), CH_4 dissolved in the effluent (2.5%) and VSS in sludge settlement (21%). In addition, the figures obtained suggest that anaerobic ponds operating under tropical conditions emitted a substantial amount of CH_4 and CO_2 compared to those reported under Mediterranean and subtropical climatic conditions and from these results, it is clear that the highest temperature leads to the highest GHG emissions. This study also corroborated that the organic loading rate (OLR) and COD influenced the production of GHGs in anaerobic ponds. The changes in COD could explain the 64% of CH_4 and CO_2 emissions produced ($p<0.05$). Further, the biogas production rate was limited by the excessive sludge accumulation in the anaerobic pond studied.

As a consequence, the CH_4 and CO_2 emissions were lower in the AP outlet than the other AP zones i.e. the entrance or central zone. Overall, these findings suggest that maintenance and design issues such as preventing sludge accumulation, capturing biogas, and reducing dissolved methane all decrease the risk of the greenhouse gases produced in anaerobic ponds being released into the environment.

The influence of the photoperiod and wastewater characteristics on greenhouse gas emissions such as CH_4, CO_2 and N_2O was studied in two pilot-scale secondary facultative ponds (Chapter 4): an algae facultative pond (AFP) and a duckweed-based pond (DBP). The results showed that under daytime conditions in the AFP median emissions were -232 mg CO_2 $m^{-2}.d^{-1}$, 9.9 mg CH_4. m^{-2} d^{-1}, and 6.9 mg N_2O m^{-2} d^{-1}, and in the DBP median emissions were -1,654.5 mg CO_2 m^{-2} d^{-1}, 1.4 mg CH_4 m^{-2} d^{-1}, and 8.5 mg N_2O m^{-2} $d,^{-1}$respectively. During night-time conditions the AFP median emissions were 3,949.9 mg CO_2 $m^{-2}.d^{-1}$, 12.7 mg CH_4, $m^{-2}.d^{-1}$, and 5.5 mg N_2O $m^{-2}.d^{-1}$ whereas the DBP median emissions were 5,116 mg CO_2 $m^{-2}.d^{-1}$, 195.2 mg CH_4 $m^{-2}.d^{-1}$, and 2 mg N_2O $m^{-2}.d^{-1}$, respectively. These figures suggest that there were significant differences between CO_2 emissions measured during daytime and night-time periods ($p<0.05$), signifying a sink-like behaviour for both the AFP and DBP in the presence of solar light, which indicates the influence of photosynthesis in the CO_2 emissions. However, once data measured during the daytime were averaged together with night-time data, the median emissions for the AFP were 1,566.8 mg CO_2 $m^{-2}.d^{-1}$, 72.1 mg CH_4 $m^{-2}.d^{-1}$, and 9.5 mg $N_2O.m^{-2}.d^{-1}$ whilst for the DBP they were 3,016.9 mg CO_2 $m^{-2}.d^{-1}$, 178.9 mg CH_4 $m^{-2}.d^{-1}$, and 8.6 mg $N_2O.m^{-2}.d^{-1}$. According to the compound average (daytime and night-time), both the AFP and DBP systems might be considered as net sources of GHG. Other findings from Chapter 4 are that the density of floating duckweed in DBPs may function as a physical barrier for gas transfer and that methane dissolved in the influent contributes in a large proportion to GHG emissions released from secondary stabilization ponds e.g. AFP and DBP.

With the differences in GHG emissions related to the photoperiod obtained previously in Chapter 4, full-scale experiments were designed to enhance understanding of daytime and night-time dynamics of CH_4, CO_2 and N_2O levels in an algal facultative pond (AFP) under tropical conditions (Chapter 5). This AFP was operated to a higher organic load and deeper than pilot-scale AFPs previously studied, which probably explained the higher GHG emissions obtained in the full-scale study. The results showed that the AFP studied was a net source of CH_4 during both daytime ($2,466.8\pm989.8$ mg CH_4 $m^{-2}.d^{-1}$) and nigh-time ($2,254\pm1,152.5$ mg CH_4 $m^{-2}.d^{-1}$). The variations in CH_4 emissions were influenced by environmental factors and physicochemical parameters such as ambient temperature and total nitrogen ($r^2=0.52$; $p<0.05$).

Probably the high concentration of DO observed in the AFP studied led to a slightly methane oxidation in the water column while low DO (1.4 mg.L^{-1} O$_2$) relates to negative ORP values (-260mV) which favoured methane production and a subsequent release into the atmosphere. For CO$_2$ emissions, a heavy influence of the photoperiod was observed. During the daytime the AFP served as a CO$_2$ sink (-743±847.5 mg CO$_2$ m^{-2} d^{-1}) while at night-time it served as a CO$_2$ source (2,497±1,334.8 mg CO$_2$ m^{-2} d^{-1}). However, the consumption of CO$_2$ observed during daytime was not enough to offset the CO$_2$ emissions of the AFP and this pond was a net source of CO$_2$. In addition, CO$_2$ production in the AFP was correlated positively to COD, and negatively to pH and DO. The significant difference between daytime and night-time CO$_2$ reflected changes in algal photosynthesis and heterotrophic respiration. Further, daytime N$_2$O fluxes from the AFP (-0.95±2.7 mg N$_2$O m^{-2} d^{-1}) and night-time (3.8±7 mg N$_2$O m^{-2}·d^{-1}) showed significant differences ($p<0.05$). Although N$_2$O varied over the 24-h cycle, fluxes did not show any clear pattern. According to the statistical test, the fluxes of N$_2$O in SFPs were influenced by NO$_3$$^-$ ($r^2=0.28$), TKN ($r^2=0.16$), DO ($r^2=0.14$), COD ($r^2=0.14$), ambient temperature ($r^2=0.12$), water temperature ($r^2=0.06$), and NH$_4$$^+$ ($r^2=0.005$). This suggests that N$_2$O could be produced by both nitrification and denitrification processes.

Chapter 6 assessed the dynamics of GHGs in a shallow eutrophic freshwater wetland under tropical conditions (the Sonso Lagoon). This investigation was included in this research because tropical wetlands play an important role in the global carbon cycle. Further, in Colombia many freshwater wetlands act as an indirect wastewater treatment system for domestic sewage treatment and therefore exhibit eutrophic conditions. The results indicate that the fluxes for the three gases showed a large variation ranging from consumption to emissions. CO$_2$ fluxes ranged from -22.9 to 23 g.m^{-2}.d^{-1} (median = 0.93), CH$_4$ ranged between -3.03 and 9.83 g.m^{-2}.d^{-1} (median = 0.04), and N$_2$O ranged from -15.2 to 12.6 mg N$_2$O m^{-2}.d^{-1} (median = 0.21). For the three gases studied, negative fluxes were observed mainly in the zone with vegetation dominated by floating plants i.e. *Eichhornia crassipes, Salvinia sp.*, and *Pistia stratiotes* L. However, the mean values indicated that the Sonso Lagoon was a net source of GHG. The effect of eutrophication on GHG emissions could be observed in the positive correlation found between CH$_4$ and CO$_2$ generation and COD, PO$_4$$^{-3}$ and NH$_3$-N. In addition, N$_2$O correlated positively to TN and NO$_3$$^-$N. This study demonstrates that pollution and eutrophication of natural wetlands results in net emissions of greenhouse gases into the atmosphere. As long as anthropogenic activities grow significantly, the organic matter and nutrient concentration i.e. N, P in freshwater wetlands will increase. This results in severe alterations of the water quality and function of these ecosystems i.e. leading to acceleration of the exchange of greenhouse gases between freshwater wetlands and the atmosphere.

Implications of EWWT on GHG emissions

This study has shown that EWWT are in general a source of GHG. Based on GWP, as was expected, the highest GWP corresponds to the anaerobic pond investigated in Chapter 3. This was attributable to high proportions of methane emitted by the anaerobic pond, which is released into the atmosphere. GWP from the full-scale anaerobic pond was higher than the AFP, the DBP and the Sonso Lagoon by 22, 184 and 33 fold, respectively. In addition, based on the organic matter removed and using the GWP showed in Table 7.2, it was found that the CO_{2-eq} in the EWWT studied was between 1.5-2.18 $kgCO_{2-eq}.kgCOD_{removed}^{-1}$. These values were higher than reported in conventional aerobic/anaerobic full treatment systems that were between 0.91 and 1.04 kg $CO_{2eq}.kgCOD_{removed}^{-1}$. Therefore, EWWT can significantly contribute to the total greenhouse gas footprint.

The values of GWP indicate the large potential of EWWT to contribute to global warming and demonstrate the necessity to improve these systems from the perspective of preventing their GHG emissions. The capture and utilization of methane emitted into atmosphere is essential to maintain low greenhouse gas production in anaerobic ponds. However, also the dissolved methane should be captured because this represents approximately 3% (Chapter 3) of COD which is equivalent to over 1.2 t CO_{2eq} . d^{-1}. This capture can be carried out by air-stripping and using off-gas in methane in energy recovery systems. Regarding secondary facultative ponds, it is clear that under tropical conditions the long photoperiod contributes to low CO_2 and CH_4. The algae and duckweed are efficient photosynthetic organisms that consume CO_2 and also provide climate change mitigation (Chapter 4 and Chapter 5). Likewise, the over-saturation of oxygen by algae photosynthesis in facultative ponds could favour biochemical methane oxidation, reducing its carbon footprint. A possible alternative to increase the potential capture of GHG in AFPs is using a hybrid process for microalgae production along with the wastewater treatment. This involves changes in the design of AFPs such as a configuration to raceway ponds, which are effective for the production of microalgae. This harvest of microalgae allows a greater capture of CO_2 and the generation of biomass that can be used in the production of biofuels and wastewater treatment.

Another important finding of this research is related to the approaches used to estimate the emissions of GHG from wastewater treatment. The IPCC has developed a method for estimating wastewater treatment emissions on a national scale. Their protocol suggests calculating emissions multiplying metrics of activity in wastewater by emission factors (EFs): the amount of GHG emitted per unit of activity. Although this methodology is quite simple, in this research it was observed that it is not the most appropriate method for estimating emissions at a particular wastewater facility. In general, the Tier 1 IPCC methodology appears to overestimate and underestimate CH_4 and N_2O emissions, respectively. This is because the emission factor is based on ideal conditions and it does not take into account that GHG emissions from wastewater depend on a number of significant influencing variables, such as carbon substrate availability, dissolved oxygen concentration and the presence of potentially inhibitory intermediates. Thus, the proposed approach for estimating GHG is to conduct site-specific source testing, followed by mass balance and modelling to validate emission factors for an area or country. This approach can contribute to reducing the knowledge gap and the uncertainties with respect to direct emissions, indirect emissions and availability and quality of data from the wastewater sector

Samenvatting

Tegenwoordig is er wereldwijde bezorgdheid over de gevolgen van de opwarming van de aarde en de klimaatverandering voor het milieu en bijgevolg ook voor de mensen die op onze planeet leven. Overmatige uitstoot van broeikasgassen - voornamelijk CO_2, CH_4 en N_2O - afkomstig van antropogene bronnen zijn de belangrijkste oorzaken van de opwarming van de aarde en de klimaatverandering. Daarom is de identificatie en kwantificering van alle bronnen, zowel natuurlijke als antropogene, een belangrijke uitdaging geworden voor wetenschappers en beleidsmakers die werken aan klimaatverandering of opwarming van de aarde, wat nodig is voor het ontwikkelen van strategieën om de snelheid van toename van de uitstoot van broeikasgassen in de atmosfeer.

Afvalwaterzuiveringssystemen zijn gebruikt voor de verwijdering van organische koolstof, nutriënten en pathogene micro-organismen uit afvalwater. Bij de behandeling van afvalwater worden organische stoffen en voedingsstoffen verwijderd wanneer ze via microbiële processen worden omgezet in gasvormige verbindingen die in de atmosfeer terechtkomen. Sommige van deze gassen, zoals CO_2, CH_4 en N_2O, kunnen echter schadelijke effecten hebben in de atmosfeer omdat ze fungeren als broeikasgassen en opwarming van de aarde veroorzaken.

Broeikasgasemissies van de afvalwaterzuiveringssector vertegenwoordigen 3-4% van de totale uitstoot van broeikasgassen. De bijdrage van de afvalwatersector aan de uitstoot van broeikasgassen kan echter worden onderschat. Dit komt door grote onzekerheden met betrekking tot directe emissies, indirecte emissies en de beschikbaarheid en kwaliteit van jaargegevens voor de afvalwatersector. Wereldwijd zijn de CH_4- en N_2O-emissies van afvalwater tussen 1990 en 2025 toegenomen van respectievelijk 352 tot 477 MtCO2eq en van 82 tot 100 MtCO2eq. Deze groei van de uitstoot van broeikasgassen komt van ontwikkelingslanden in Oost- en Zuid-Azië, het Midden-Oosten, het Caribisch gebied en Midden- en Zuid-Amerika, voornamelijk als gevolg van de bevolkingsgroei. Zolang de bevolking aanzienlijk groeit zonder grootschalige vooruitgang in de behandeling van afvalwater, zullen deze gebieden een grote invloed blijven hebben op de opwaartse trend in de uitstoot van broeikasgassen.

Ecotechnologische systemen voor afvalwaterbehandeling (EWWT) combineren ecologische principes van natuurlijke systemen met technische principes om de verwijdering van organische koolstof, voedingsstoffen en pathogene micro-organismen uit afvalwater te verbeteren. De belangrijkste EWWT systemen zijn anaerobe vijvers, facultatieve vijvers, vijvers met eendenkroos en aangelegde wetlands. EWWT systemen zijn voornamelijk op zonne-energie gebaseerd, waardoor de afhankelijkheid van externe energie minimaal is en daarom worden ze beschouwd als duurzame technologieën. EWWT systemen zijn echter een belangrijke bron van broeikasgassen zoals CO_2, CH_4 en N_2O en dragen waarschijnlijk bij aan het broeikaseffect.

Tot nu toe is er weinig informatie beschikbaar over de productie of het verbruik van broeikasgassen bij EWWT systemen die onder tropische omstandigheden worden gebruikt. Dit doctoraatsonderzoek kwantificeert de CO_2-, CH_4- en N_2O-fluxen gegenereerd in drie EWWT systemen: anaerobe vijvers, facultatieve vijvers en vijvers met eendenkroos. Ook werd de flux van broeikasgassen gemeten uit een zoetwater-wetland, verstoord door antropogene activiteiten zoals afvoer van afvalwater en voedingsstoffen uit landbouwafvoer. Dit onderzoek omvatte niet geconstrueerde wetlands omdat, volgens de literatuurstudie, deze grote aandacht hebben gekregen met betrekking tot broeikasgas-metingen. Verder bestudeerde dit proefschrift de invloed van omgevingsfactoren zoals temperatuur, pH, DO en nutriënten op broeikasgasemissies geproduceerd in EWWT-systemen onder tropische omstandigheden. De methodologie die in het onderzoek is toegepast omvat zowel experimenten op laboratoriumschaal als in echte zuiveringen. Alle studies werden uitgevoerd met behulp van het afvalwater dat werd geproduceerd in een klein stadje in Colombia, genaamd Ginebra, en geen van de studies gebruikte synthetisch afvalwater.

Het eerste deel van het onderzoek in dit proefschrift (hoofdstuk 2) was gericht op de aanpassing en beoordeling van een meettechniek om de broeikasgasemissies in EWWT-systemen te schatten. De techniek van de statische kamer is gekozen omdat deze de meest bruikbare en gerapporteerde techniek is voor het meten van broeikasgassen uit afvalwaterinstallaties, aquatische ecosystemen en bodems. De verkregen resultaten suggereren dat gesloten statische kamers een goede analytische methode zijn om de broeikasgasemissies van afvalwaterstabilisatievijvers te schatten. Er is echter een beperking aan het gebruik van gesloten statische kamers, want niet alle gegevens voor gasconcentraties gemeten in de headspace van een kamer kunnen worden gebruikt om de flux te schatten omdat er gradiëntconcentratiecurves met niet-plausibele en fysieke verklaringen voorkomen. Op basis van de totale dataset (n=47), was het percentage aanvaarde curven respectievelijk 94, 87 en 73% voor CH_4, CO_2 en N_2O. Bovendien toonden de statistische analyses aan dat alleen het nemen van lineaire regressie vaak ongeschikt was voor de bepaling van de broeikasgas-flux uit stabilisatiebassins door de gesloten statische kamertechniek. Uit dit werk wordt het duidelijk dat wanneer $R^2_{adj-non-lin} > R^2_{adj-lin}$ de toepassing van lineaire regressiemodellen niet wordt aanbevolen en dit leidt tot een onderschatting van de broeikasgas-flux met 10 tot 50%. Daarom is de belangrijkste conclusie van hoofdstuk 2 dat onjuist gebruik van de gebruikelijke R2-parameter en alleen het lineaire regressiemodel voor het schatten van de fluxen kan leiden tot een ernstige onderschatting van de werkelijke bijdrage van broeikasgasemissies van afvalwater.

Het uitvoeren van een volledig onderzoek naar een anaerobe vijver waarin huishoudelijk afvalwater werd behandeld, was een poging om prestatiegegevens te geven over broeikasgassen zoals CO_2 en CH_4, en N_2O onder tropische omstandigheden (hoofdstuk 3). CH_4-emissies varieerden van 13,4 tot 178,7 $Lm^{-2}.d^{-1}$, CO_2 van 9.3 tot 130.5 $Lm^{-2}.d^{-1}$, terwijl de N_2O-emissies varieerden van 0.0016 tot 0.013 $Lm^{-2}.d^{-1}$. Volgens de gemiddelde fluxen waren de emissies in de atmosfeer voor CH_4 en CO_2 respectievelijk 0,24 m^3 CH_4 /kg CZV_{rem} en 0,18 m^3 CO_2 /kg CZV_{rem}. Verder gaf een CZV-massabalansberekening aan dat 37% van de influent CZV werd omgezet in CH_4 en 36% de anaerobe vijver met het effluent verliet. De rest van de CZV zat in vluchtige vaste stoffen (3,5%), CH_4 opgelost in het effluent (2,5%) en VSS in slibbereiding (21%). Bovendien geven de verkregen cijfers aan dat anaerobe vijvers die onder tropische omstandigheden werken een aanzienlijke hoeveelheid CH_4 en CO_2 uitstootten in vergelijking met de vijvers onder mediterrane en subtropische klimatologische omstandigheden. Uit deze resultaten blijkt duidelijk dat de hoogste temperatuur leidt tot de hoogste uitstoot van broeikasgassen. Deze studie bevestigde ook dat de organische belasting en CZV de productie van broeikasgassen in anaerobe vijvers beïnvloeden. De veranderingen in CZV kunnen 64% van de geproduceerde CH_4- en CO_2-emissies verklaren (p <0,05). Verder werd de biogasproductiesnelheid beperkt door de overmatige accumulatie van slib in de bestudeerde anaërobe vijver. Dientengevolge waren de CH_4- en CO_2-emissies lager in de AP-uitlaat dan de andere AP-zones, d.w.z. de ingangs- of centrale zone. Over het algemeen suggereren deze bevindingen dat onderhouds- en ontwerpkwesties zoals het voorkomen van slibophoping, het opvangen van biogas en het verminderen van opgelost methaan alle het risico verminderen dat de broeikasgassen die vrijkomen in anaerobe vijvers in het milieu terechtkomen.

De invloed van de fotoperiode en afvalwatereigenschappen op broeikasgasemissies zoals CH_4, CO_2 en N_2O werd bestudeerd in twee proefopstellingen van secundaire facultatieve vijvers (hoofdstuk 4): een vijver met algen (AFP) en een vijver met eendenkroos (DBP). Gedurende de dag waren de mediane emissies in de AFP -232 mg CO_2 $m^{-2}.d^{-1}$, 9.9 mg CH_4 $m^{-2}.d^{-1}$ en 6,9 mg N_2O $m^{-2}.d^{-1}$ waren, en in de DBP -1654,5 mg CO_2 $m^{-2}.d^{-1}$, 1,4 mg CH_4 $m^{-2}.d^{-1}$ en 8,5 mg N_2O $m^{-2}.d^{-1}$ respectievelijk. Tijdens de nacht bedroegen de mediane emissies van de AFP 3949,9 mg CO_2 $m^{-2}.d^{-1}$, 12,7 mg CH_4. $m^{-2}.d^{-1}$, en 5,5 mg N_2O $m^{-2}.d^{-1}$, terwijl in de DBP de mediane emissies 5166 mg CO_2 $m^{-2}.d^{-1}$, 195,2 mg CH_4 $m^{-2}.d^{-1}$ en 2 mg N_2O $m^{-2}.d^{-1}$ waren, respectievelijk. Deze resultaten suggereren significante verschillen tussen CO_2-emissies gemeten overdag en 's nachts (p <0,05), maar in beide gevallen fungeren zowel AFP als DBP als opslag wat de invloed van fotosynthese op de CO_2 uitstoot laat zien. Wanneer de dag- en nachtresultaten werden gecombineerd waren de mediane emissies voor de AFP 1566,8 mg CO_2 $m^{-2}.d^{-1}$, 72.1 mg CH_4 $m^{-2}.d^{-1}$ en 9.5 mg $N_2O.m^{-2}.d^{-1}$ en voor de DBP 3016,9 mg CO_2 $m^{-2}.d^{-1}$, 178,9 mg CH_4 $m^{-2}.d^{-1}$ en 8.6 mg $N_2O.m^{-2}.d^{-1}$. Volgens het samengestelde gemiddelde (overdag en 's nachts), kunnen zowel de AFP- als de DBP-systemen worden beschouwd als netto bronnen van broeikasgassen. Andere bevindingen uit hoofdstuk 4 zijn dat de dichtheid van drijvend kroos in DBP's kan functioneren als een fysieke barrière voor gasoverdracht en dat methaan opgelost in het influent in grote mate bijdraagt aan broeikasgasemissies die vrijkomen uit secundaire stabilisatievijvers, b.v. AFP en DBP.

Met de verschillen in broeikasgasemissies gerelateerd aan de eerder in hoofdstuk 4 verkregen fotoperiode, werden experimenten op ware grootte ontworpen om het begrip van de dag- en nachtdynamiek van CH_4-, CO_2- en N_2O-niveaus in een facultatieve algenvijver (AFP) onder tropische omstandigheden te verbeteren (hoofdstuk 5). Deze AFP werkte met een hogere organische belasting en was dieper dan de eerder onderzochte pilot-schaal AFP's, wat waarschijnlijk de hogere broeikasgasemissies verklaarde. De resultaten toonden aan dat de onderzochte AFP een netto bron van CH_4 was zowel overdag (2466,8 $\pm$ 989,8 mg CH_4 $m^{-2}.d^{-1}$) en 's nachts (2254 $\pm$ 1152,5 mg CH_4 $m^{-2}.d^{-1}$). De variaties in CH_4-emissies werden beïnvloed door omgevingsfactoren en fysisch-chemische parameters zoals omgevingstemperatuur en totale stikstof (r^2 = 0,52; p <0,05). Waarschijnlijk leidde de hoge DO-concentratie die in de onderzochte AFP werd waargenomen tot een beperkte methaanoxidatie in de waterkolom, terwijl een lage DO (1,4 mg.L^{-1} O_2) betrekking had op negatieve ORP-waarden (-260mV) die methaanproductie en een daaropvolgende afgifte begunstigden. Voor CO_2-emissies werd een grote invloed van de fotoperiode waargenomen. Overdag diende de AFP als een CO_2-put (-743 $\pm$ 847,5 mg CO_2 $m^{-2}.d^{-1}$) terwijl het 's nachts diende als een CO_2-bron (2497 $\pm$ 1334,8 mg CO_2 $m^{-2}.d^{-1}$). Het verbruik van CO_2 dat overdag werd waargenomen was echter niet voldoende om de CO_2-uitstoot van de AFP te compenseren en deze vijver was een netto-bron van CO_2. Bovendien was de CO_2-productie in de AFP positief gecorreleerd met CZV en negatief met de pH en DO. Het significante verschil tussen CO_2 tijdens dag en nacht weerspiegelde veranderingen in fotosynthese van algen en heterotrofe ademhaling. Verder vertoonde overdag N_2O uit de AFP (-0,95 $\pm$ 2,7 mg N_2O $m^{-2}.d^{-1}$) en 's nachts (3,8 $\pm$ 7 mg N_2O $m^{-2}.d^{-1}$) significante verschillen (p <0,05). Hoewel N_2O gedurende de cyclus van 24 uur varieerde, vertoonden fluxen geen duidelijk patroon. Volgens de statistische test werden de fluxen van N_2O in SFP's beïnvloed door NO^{3-} (r^2 = 0.28), TKN (r^2 = 0.16), DO (r^2 = 0.14), CZV (r^2 = 0.14), omgevingstemperatuur (r^2 = 0.12), watertemperatuur (r^2 = 0,06) en NH_4^+ (r^2 = 0,005). Dit suggereert dat N_2O zowel door nitrificatie als door denitrificatie kan worden geproduceerd.

Hoofdstuk 6 beschrijft de dynamiek van broeikasgassen in een ondiep eutroof zoetwater-moerasgebied onder tropische omstandigheden (de Sonso-lagune). Dit onderzoek is in dit proefschrift opgenomen omdat tropische wetlands een belangrijke rol spelen in de wereldwijde koolstofcyclus en in Colombia fungeren veel zoetwaterrijke gebieden als een indirect afvalwaterbehandelingssysteem voor huishoudelijk afvalwater en ze zijn daarom eutroof. De resultaten geven aan dat de fluxen voor de drie gassen een grote variatie vertoonden, van consumptie tot emissies. CO_2-fluxen varieerden van -22,9 tot 23 $gm^{-2}.d^{-1}$ (mediaan = 0.93), CH_4 varieerde van -3.03 tot 9.83 $gm^{-2}.d^{-1}$ (mediaan = 0.04), en N_2O varieerde van -15.2 tot 12.6 mg N_2O $m^{-2}.d^{-1}$ (mediaan = 0.21). Voor de drie onderzochte gassen werden negatieve fluxen waargenomen, voornamelijk in de zone met vegetatie gedomineerd door drijvende planten, d.w.z. *Eichhornia crassipes, Salvinia* sp. en *Pistia stratiotes*. De gemiddelde waarden gaven echter aan dat de Sonso-lagune een netto bron van broeikasgassen was. Het effect van eutrofiëring op de uitstoot van broeikasgassen kon worden waargenomen in de positieve correlatie tussen CH_4 en CO_2-vorming en CZV, PO_4^{-3} en NH_3-N. Bovendien correleerde N_2O positief met TN en NO_3-N. Deze studie toont aan dat vervuiling en eutrofiëring van natuurlijke wetlands resulteert in netto emissies van broeikasgassen in de atmosfeer. Zolang de antropogene activiteiten aanzienlijk groeien, zullen de organische stof en nutriëntenconcentratie (N, P) in zoetwaterrijke gebieden toenemen. Dit resulteert in ernstige veranderingen van de waterkwaliteit en functie van deze ecosystemen en leidt tot versnelling van de uitwisseling van broeikasgassen tussen zoetwatermoeraslanden en de atmosfeer.

Implicaties van ecotechnologische systemen op de uitstoot van broeikasgassen

Deze studie heeft aangetoond dat ecotechnologische systemen in het algemeen bronnen van broeikasgassen zijn. Het hoogste broeikasgasemissie potentieel komt uit de anaerobe vijver die in hoofdstuk 3 werd onderzocht. Dit was te wijten aan het hoge gehalte methaan dat werd uitgestoten door de anaerobe vijver. Het broeikasgasemissie potentieel van de anaerobe vijver was respectievelijk 22, 184 en 33 keer hoger dan van de AFP, de DBP en de Sonso-lagune. Bovendien werd, op basis van de verwijderde organische stof en met behulp van de broeikasgasemissie potentieel waarden in Tabel 7.2, gevonden dat de CO_2-eq in de onderzochte ecotechnologische systemen tussen 1,5 en 2,18 kg CO_2-eq.kgCZV$_{verwijderd}^{-1}$ lag. Deze waarden waren hoger dan gerapporteerd in conventionele aerobe/anaerobe volledige behandelingssystemen die tussen 0,91 en 1,04 kg CO_2eq.kgCOD$_{verwijderd}^{-1}$ waren. Daarom kunnen ecotechnologische systemen aanzienlijk bijdragen aan de totale broeikasgasvoetafdruk.

De broeikasgasemissie potentiëlen van ecotechnologische systemen kunnen in hoge mate bijdragen aan het broeikaseffect en tonen de noodzaak aan om deze systemen te verbeteren om hun broeikasgasemissies te voorkomen. Het wegvangen en het hergebruik van methaan zijn essentieel om de lage broeikasgasproductie in anaerobe vijvers te behouden. Het opgeloste methaan moet echter ook worden opgevangen omdat dit ongeveer 3% (hoofdstuk 3) CZV vertegenwoordigt, wat overeenkomt met meer dan 1,2 t CO_2eq. d^{-1}. Dit kan gebeuren door strippen met lucht in energieterugwinningssystemen. Met betrekking tot secundaire facultatieve vijvers is het duidelijk dat de lange lichtperiode onder tropische omstandigheden bijdraagt aan een lage CO_2-uitstoot en CH_4. De algen en het eendenkroos zijn efficiënte fotosynthetische organismen die CO_2 consumeren en ook zorgen voor beperking van de klimaatverandering (hoofdstuk 4 en hoofdstuk 5). Evenzo kan de oververzadiging van zuurstof door fotosynthese door algen in facultatieve vijvers biochemische methaanoxidatie bevorderen, waardoor de koolstofvoetafdruk ervan afneemt. Een mogelijk alternatief om de potentiële opname van broeikasgassen in AFP's te vergroten, is het gebruik van een hybride proces voor de productie van microalgen, samen met de behandeling van afvalwater. Dit omvat wijzigingen in het ontwerp van AFP's, zoals een configuratie voor kanaalvijvers, die effectief zijn voor de productie van microalgen. Het oogsten van microalgen maakt een grotere afvang van CO_2 en de productie van biomassa mogelijk die kan worden gebruikt bij de productie van biobrandstoffen en afvalwaterzuivering.

Een andere belangrijke uitkomst van dit onderzoek houdt verband met de benaderingen die worden gebruikt om de emissies van broeikasgassen door afvalwaterzuiveringen te schatten.

Het IPCC heeft een methode ontwikkeld voor het schatten van emissies van afvalwaterzuivering op nationale schaal. Hun protocol suggereert de gemeten emissies te vermenigvuldigen met emissiefactoren: de hoeveelheid broeikasgassen die wordt uitgestoten per eenheid van activiteit. Hoewel deze methode vrij eenvoudig is, is in dit onderzoek gevonden dat dit niet de meest geschikte methode is om de emissies in een bepaalde afvalwaterinstallatie te schatten. Over het algemeen lijkt de IPCC-methode van Tier 1 de CH_4- en N_2O-emissies respectievelijk te overschatten en te onderschatten. Dit komt omdat de emissiefactor gebaseerd is op ideale omstandigheden en er geen rekening mee wordt gehouden dat de broeikasgasemissies van afvalwater afhankelijk zijn van een aantal significante beïnvloedende variabelen, zoals beschikbaarheid van koolstofsubstraat, concentratie van opgeloste zuurstof en de aanwezigheid van potentieel remmende tussenproducten. De voorgestelde aanpak voor het schatten van broeikasgassen is dus om site-specifieke brontests uit te voeren, gevolgd door massabalans en modellering om emissiefactoren voor een gebied of land te valideren. Deze aanpak kan bijdragen aan het verkleinen van de kenniskloof en de onzekerheden met betrekking tot directe emissies, indirecte emissies en beschikbaarheid en kwaliteit van gegevens uit de afvalwatersector

Curriculum Vitae

Juan Pablo Silva Vinasco was born on 27th of March 1964 in Cali, Colombia. After, completing high school, he proceeded to Universidad del Valle in 1982 where he obtained the BSc. in Chemical Engineering. After his graduation he worked in the industrial sector as process engineer until 1996 when he enrolled in the MSc in Sanitary and Environmental Engineering Program of Universidad del Valle.

After his postgraduate studies in 1998 he was employed in the Universidad del Valle where he currently works as associate professor in the School of Natural Resources and Environment of the Engineering Faculty. He teaches both undergraduate and post-graduated courses in environmental processes, air pollution, and industrial ecology. His current research interest include greenhouse gas measurements in aquatic and terrestrial ecosystems, biological waste gas treatment, air pollution modelling, life cycle assessment.

In 2007 he began his Ph.D. study at UNESCO-IHE Institute for Water Education supported by SWITCH project (Sustainable Urban Water Management Improves Tomorrow's City's Health). During his PhD. study he supervised several MSc projects in Universidad del Valle.

Publications of last five years

1.FRANCISCO CAICEDO, JOSÉ M. ESTRADA, JUAN P SILVA. RAÚL MUÑOZ, RAQUEL LEBRERO (2018). "Effect of packing material configuration and liquid recirculation rate on the performance of a biotrickling filter treating VOCs". Journal of chemical technology and biotechnology. doi:10.1002/jctb.5573

2. YELITZA ZORRILLA, JUAN PABLO SILVA VINASCO, VALENTINA PRADO, PABLO MANYOMA. "Evaluation of water use in food SMEs: case study of a poultry in Colombia". DYNA, 85(206), pp. 226-235, September, 2018. DOI: https://doi.org/10.15446/dyna.v85n206.68809

3. JUAN PABLO SILVA VINASCO, ANA PAOLA LASSO PALACIOS, MIGUEL RICARDO PENA VARON, HENK LUBBERDING, HUUB GIJZEN (2015), "Biases in greenhouse gases static chambers measurements in stabilization ponds: Comparison of flux estimation using linear and non-linear models". Atmospheric Environment ISSN: 1352- 2310 ed: Pergamon v.109 fasc.1 p.130 - 138 (2015)

4. JUAN PABLO SILVA VINASCO, JOSE LUIS RUIZ MARTINEZ, MIGUEL RICARDO PENA VARON, HENK LUBBERDING, HUUB GIJZEN (2012), "Influence of photoperiod on carbon dioxide and methane emissions from two pilots-scale stabilization ponds" Water Science and Technology ISSN: 0273-1223 ed: Pergamon v.66 fasc.9 p.1930 - 1940 (2012)